B

BOILER CONTROL SYSTEMS

David Lindsley
Consultant in Control and Instrumentation

McGRAW-HILL BOOK COMPANY

London · New York · St Louis · San Francisco · Auckland · Bogotá
Caracas · Hamburg · Lisbon · Madrid · Mexico · Milan · Montreal
New Delhi · Panama · Paris · San Juan · São Paulo · Singapore
Sydney · Tokyo · Toronto

Published by
McGRAW-HILL Book Company (Europe)
Shoppenhangers Road, Maidenhead, Berkshire, SL6 2QL, England
Telephone 0628 23432
Fax 0628 770224

Published by
McGRAW-HILL Book Company Europe

British Library Cataloguing in Publication Data
Lindsley, David
 Boiler control systems.
 1. Great Britain. Power stations. Control systems
 I. Title
 621.312132

Library of Congress Cataloging-in-Publication Data
Lindsley, David
 Boiler control systems/David Lindsley
  p.   cm.
 Includes bibliographical references and index.
 ISBN 0-07-707374-6
 1. Steam-boilers – Automatic control. I Title.
TJ288.L48   1991
621.1′83 – dc20

91-8444
CIP

Copyright © 1991 McGraw-Hill International (UK) Limited. All rights reserved. No part of this publication may be reproduced, stored in a retrieval system, or transmitted, in any form or by any means, electronic, mechanical, photocopying, recording, or otherwise, without the prior permission of McGraw-Hill International (UK) Limited.

1234 CUP 9321

Typeset by Paston Press, Loddon, Norfolk
Printed and bound in Great Britain at the University Press, Cambridge

CONTENTS

ACKNOWLEDGEMENTS	ix
CHAPTER 1 INTRODUCTION	**1**
1.1 The background, purpose and scope of this book	1
1.2 Distinguishing aspects of boiler control technology	2
1.3 The need to understand the whole closed loop	3
1.4 The rapid evolution of electronics	4
1.5 Users' preferences	4
1.6 Units of measurement	5
CHAPTER 2 POWER-STATION PLANT	**6**
2.1 Boiler/turbine units and the range system	6
2.2 Some characteristics of steam	7
2.3 An outline of boilers	9
2.3.1 Feed-water systems	13
2.3.2 The steam circuits	14
2.3.3 The combustion process	16
2.3.4 The waste products of the combustion process	17
2.3.5 Draught plant	19
2.3.6 Fuel systems	22
2.3.7 Treatment of flue gases	27
2.4 The steam turbine and the alternator	29
2.4.1 The steam turbine	29
2.4.2 Condensate systems	31
2.4.3 The alternator	31
2.5 Once-through boilers	32
2.6 Feed-water conditioning	33
2.7 Turbine bypass valves	33
2.8 Cyclic life-expenditure curves of turbines	34
2.9 Combined-cycle plants	35
2.10 Co-generation	36
Notes	37

CHAPTER 3 MODULATING CONTROL SYSTEMS   38
3.1 The master controller   39
    3.1.1 'Boiler-following' mode of operation   40
    3.1.2 'Turbine-following' mode of operation   41
    3.1.3 'Sliding-pressure' mode of operation   43
3.2 Combustion control   51
    3.2.1 The fuel/air ratio   52
    3.2.2 A simple parallel combustion control system   58
    3.2.3 'Cross-limited' combustion control   60
    3.2.4 Control of larger systems   62
    3.2.5 Combustion control in fluidized-bed boilers   63
3.3 Pulverizer control   63
    3.3.1 The control of pressurized vertical-spindle pulverizers   64
    3.3.2 The control of suction-type vertical-spindle pulverizers   66
    3.3.3 The control of horizontal-tube pulverizers   66
3.4 Mixed-fuel firing   67
    3.4.1 Problems associated with the control of multiple pulverizers   68
    3.4.2 A typical pulverizer control system   70
3.5 Control of air flow   73
    3.5.1 Parallel operation of FD fans   73
    3.5.2 The air-flow demand signal   75
    3.5.3 Air-flow control in fluidized-bed boilers   77
3.6 Furnace pressure control   78
    3.6.1 Furnace pressure control in boilers with FGD   81
3.7 Feed-water control   81
    3.7.1 Level control in drum boilers   83
    3.7.2 Feed-water control in once-through boilers   87
3.8 Steam temperature control   96
    3.8.1 Superheated steam temperature control   97
    3.8.2 Multistage superheaters   100
    3.8.3 Steam temperature control in fluidized-bed boilers   101
    3.8.4 Reheated steam temperature control   101
3.9 The control of combined-cycle plants   102
3.10 The control of co-generation plants   105
    Notes   107

CHAPTER 4 ANCILLARY SYSTEMS   109
4.1 Turbine bypass control   109
4.2 HP bypass   110
4.3 LP bypass   112
    Note   112

CHAPTER 5	**SEQUENCES AND INTERLOCKS**	**113**
5.1	Interlock systems	114
5.2	Sequence systems	116
	5.2.1 A typical sequence—start-up of an FD fan	116
5.3	Burner management systems	118
CHAPTER 6	**EQUIPMENT CONSIDERATIONS**	**123**
6.1	Manual operation	124
6.2	Environmental considerations	128
	6.2.1 Power supplies	128
	6.2.2 Electromagnetic compatibility	130
	6.2.3 Temperature	131
	6.2.4 Vibration	132
	6.2.5 Impact, shock, rough usage and earthquake	132
	6.2.6 Dust and dirt	132
	6.2.7 Cabling, shielding and grounding	132
6.3	The need for comprehensive specifications	133
6.4	Performance testing	136
6.5	Control-room layout	137
	Notes	139

APPENDIX 1	**CONVERSION FACTORS AND OTHER USEFUL INFORMATION**	**140**
APPENDIX 2	**SYMBOLS USED ON CONTROL DIAGRAMS**	**145**
APPENDIX 3	**MULTILINGUAL GLOSSARY OF TERMS**	**146**
INDEX		**154**

ACKNOWLEDGEMENTS

As a young electronics student, part of the industrial training section of my 'sandwich course' was at Kingston 'B' Power Station, to the south-west of London. At that time, I regarded this as a necessary but slightly unwelcome intrusion of heavy electrical engineering into my world of thermionic valves and microamps.

On my first day on site, I realized that all the subjects I had so far considered to be peripheral courses at College (such as thermodynamics and engineering chemistry) were real, living and important entities. Moreover, I discovered that the world of noise and heat in which I suddenly found myself was vastly more exciting than the well ordered electronics laboratories that I had at that time planned to enter.

The staff who took time to train me in those days gave me my first grounding in power-plant engineering. They frightened the pants off me by making me climb the boilers to service furnace tappings, and by getting me to crawl into the narrow plenum chamber between the top and bottom sections of the chain grate to free jammed dampers, but I owe them a lot.

As a result of my experience with the six 30 MW chain-grate-fired Stirling boilers at Kingston 'B', I decided that my career would lie with the power generation industry.

I can see the silent hulk of Kingston 'B' out of my offices now. By modern standards, it is tiny and archaic, and it has lain empty and quiet for around 30 years. But I owe it a lot, too. The doors it opened for me took me right round the world and brought me adventures, comradeship and sheer exhilaration, which I would never have otherwise experienced.

So, in giving thanks to those who helped me write this book, I must first of all acknowledge that 'old lady' across the river! In time it may be knocked down, and when that time comes I hope it will be converted into a modern plant, again serving the community quietly and efficiently as it once did. (What I would *not* like would be to see it flattened for some transient office development or—worse still—gutted and stuffed with alien equipment like its counterpart at Battersea!) Whatever happens, it taught me a lot and put me on the path that filled my life.

When it came to writing this book, help came from the many individuals and companies who sent me contributions and photographs and who willingly took time off to clarify some point or other. The following is a list of all those I can remember, but I am sure that there are bound to be some I forgot. In that case, I hope they will accept my apologies.

- Mr F Stalder, ASEA Brown Boveri Power Generation Ltd, Baden, Switzerland
- James R. Allen, Bailey Controls Co., Wickliffe, Ohio, USA
- Babcock Energy Ltd, Crawley, UK.
- Bristol-Babcock Ltd, Dorking and Kidderminster, UK
- Burns and McDonnell, Kansas City, Missouri, USA
- China Light and Power Co. Ltd, Hong Kong
- Krupp Atlas Electronik GmBH, Bremen, Germany
- Rob Smith, NEI Power Projects, UK
- Schlumberger Industries (Transducer Division), Farnborough, Hants, UK
- Senior Foster Wheeler Ltd, Wembley, UK
- Servomex Ltd, Crowborough, Sussex, UK
- Hr Karl-Heinz Supancic, Siemens AG, Karlsruhe, Germany
- Scottish Power, Glasgow, Scotland
- Mr Peter Cox, State Electricity Commission of Victoria, Australia
- Hartman and Braun AG

I hope that this book will be useful to many different kinds of people, whether they are students, control engineers, computer specialists or power-plant staff. If nothing else, I hope that it avoids somebody finding out—the hard way—that power-station control systems are much more complex than an outsider could possibly imagine, and that many electronic devices just fall over when they are exposed to that demanding environment. That was another lesson I learned all those years ago at Kingston 'B'!

David Lindsley
Kingston-upon-Thames
Surrey, UK

CHAPTER
ONE

INTRODUCTION

1.1 THE BACKGROUND, PURPOSE AND SCOPE OF THIS BOOK

Before the final decade of the twentieth century, people had become aware of the pervasive and damaging side-effects of their ever-growing thirst for energy—and of their profligate use of it. At this time, and perhaps unfairly, attention focused to a large extent on power generating plant, and within the industry these concerns pointed to the need for safer, more cost-effective and more environmentally benign power plant. The outcome was a steady development of improved processes, plant design and materials and—since greater efficiency and cleaner operation are assisted by better monitoring and control—of enhanced means of controlling the relevant processes.

This is the situation we live with now. Its birth coincided with the rapid evolution of the digital computer and with the explosion of technology generally. Not altogether surprisingly, this has led to some strange anomalies: on one hand a perceived need for better understanding and control of the power generation process, and on the other the growth of specialization leading to the development of allied, yet isolated, technical disciplines. The dichotomy—of computer engineers becoming involved in a complex process they often do not understand, and power engineers being expected to grasp the complex and shifting silicon sands of computers—runs the risk of producing misunderstandings and errors, which could be dangerous in any situation, let alone in a power-plant boiler, which is a large and dangerous process with little room for mistakes.

This book aims to try to assist in resolving the dilemma. It is an analysis of control technology applied to power-plant steam generating systems. It covers a wide variety of plant, including public-utility electricity generation systems and privately owned and operated power utilities on industrial sites. The many different combustion arrangements that are covered include those based on oil, gas and coal firing, and those employing conventional firing or fluidized-bed combustion. The operational regimes range from drum types to supercritical once-through units.

Although the basic operational principles and control systems of each of these types of plant have their own unique characteristics, each bears some similarity to the others, and the book therefore attempts to cover both the common principles and the special requirements.

The aim of this book is to address a broad spectrum of engineers, including plant people who have some familiarity with control, but the target readers are predominantly control or computer specialists who are familiar with the general concepts of automation, and who need to apply that knowledge to a boiler about which they actually know very little. For this reason, the operational principles of the plant have in many instances been considerably simplified (some may say they have been reduced to the ridiculous!)—although in all such situations the book goes on at least to indicate the real situation in more depth and with greater accuracy.

1.2 DISTINGUISHING ASPECTS OF BOILER CONTROL TECHNOLOGY

The subject of boiler control demands special attention because it stands apart from that of most other processes on many counts. Three of these are as follows:

1. A boiler comprises a complex series of heat exchanges, which are *interactive, multivariable* processes, demanding complex feedforward and feedback control and very special human/machine interface considerations.
2. Conventional control theory usually emphasizes the need to relate the closed-loop criteria to a mathematical model of the plant—but the mathematical definitions of combustion plant and processes are usually inadequately known, and often poorly documented.
3. For all its complexities and interactions, a boiler is only part of the power plant, and care must be taken to consider its interactions with the rest of the installation, notably the steam turbine, which is itself a special case, especially in view of the enormous progress in the size and complexity of these machines. (In fact, the interrelationship between boiler and turbine is so important that, although a detailed description of turbine characteristics and control systems is beyond

the scope of the present book, an overview of the subject is presented in Sec. 2.4.)

These three factors draw attention to the need for employing special skill and care in designing boiler control systems, but, as already mentioned, the combustion process is hazardous. The hazards are inherent in the process, and deficiencies in the control systems can damage the plant itself, injure or kill people, or damage the environment on a local, or sometimes global, scale. In this respect a boiler is no worse than, say, a chemical refinery; but it does differ in the respect that seemingly trivial disturbances can involve quite massive amounts of energy. For example, in a 660 MW boiler a change or disturbance amounting to 0.1 per cent is in reality equivalent to a sizeable 660 kW, or nearly 900 horsepower!

Because of the problems presented by plant response times, and the significance of small changes in firing rate (especially with the largest installations), it becomes important to consider very carefully indeed the design of the control system and its response time.

In spite of these very special and demanding conditions, the subject of boiler control has received only sparse coverage in textbooks—or at least in up-to-date ones—and this is something that this book aims to redress.

1.3 THE NEED TO UNDERSTAND THE WHOLE CLOSED LOOP

The growth in complexity of technology has forced on the engineering community the need to segregate disciplines and to narrow the field of each speciality to what can be reasonably assimilated.

Unfortunately, this narrow specialization often conflicts with the practicalities of power-plant applications. The performance of a control loop is governed by the characteristics of its constituent parts—the plant, the controller and the transducers—and therefore the design of effective, safe and reliable control systems demands a knowledge of all these areas. The boiler control specialist must therefore combine a thorough understanding of computers and control theory with an equal familiarity with the characteristics and the intricacies of the plant: a tall order, especially since the study of digital computers—a full-time exercise in its own right—is subject to continual and rapid changes.

This is the reason why this book aims to provide a complete picture of the plant, the control systems and the interfaces. (However, to contain the scale to a reasonable level, the architecture of the process is described in only broad-brush terms, leaving the intricacies of the constituent parts to the domain of specialist works.)

1.4 THE RAPID EVOLUTION OF ELECTRONICS

The advent of digital electronics technology has transformed the nature of practical boiler control applications out of all recognition. Engineers who were familiar with power utilities as they existed in the first half of this century would doubtless recognize most of today's major plant items such as pumps, fans, burners and turbines (although they may well be surprised at their size), and they would have few problems in identifying the control valves and actuators. But they would have great difficulty in grasping the concepts behind the 'heart and brains' of the control system—often computer-based with complex graphics displays. To such people, this part of the control equipment would seem to be completely alien, almost magical in its operation.

Although it seems unlikely that we should ever see such a complete metamorphosis as occurred over the middle of the twentieth century—when the prevalent systems changed from being essentially mechanical (pneumatic or hydraulic) to being almost completely electronic—the essential nature of control systems continues to change, and a description of control technology runs the risk of being outdated by developments in electronics and computers, and, to a lesser extent, by changes in the area of the plant.

This book is therefore directed towards defining the known and less changeable parameters: the primary plant characteristics and the established control loops. This definition is combined with surveys of how the various control techniques have evolved and where they seem to be heading, and, since it is often useful to examine previous case histories (if only to see what the advantages and disadvantages proved to be), details of several current applications are included.

1.5 USERS' PREFERENCES

Although there is a steady trend towards standardization, certain users have strong preferences for customized features. Although these are often in the field of control-room ergonomics (often called 'human factors'), examples can be found in almost every discipline. In many instances, these preferences appear to be nationalistic.

For example, in certain countries the practice is for the interfaces between the controlling and controlled part of the plant to take the form of multiphase stepper-motor drives: in other countries, analogue signals are used for this purpose.

Again, whereas in the Americas (and in countries dominated by US thinking) control consoles tend to be of sheet metal construction with individual instruments accommodated in cut-outs in a solid desk, most

European control desks are made up of mosaic tiles mounted on matrix frames.

A special consideration relates to software. At the time of writing this book, some users accept whatever software seems to be the best for each application, while others go to great lengths to specify the application software in every aspect.

The tendency is for such preferences to be most apparent in the area of new developments—with standardization and uniformity progressively taking over as the technology settles down and becomes established. It is therefore likely that the individual examples described in this book will be superseded by national and international standards—though it seems inevitable that new preferences will continue to appear and develop in specific application areas.

In one area these preferences presented special difficulties in the preparation of this book—that is, the symbols used on control diagrams. Because many nations' technologies are examined in detail here, no single national or international standard for symbolic notation proved adequate for showing the fine details of control techniques as they are applied in various countries. Therefore, in spite of the strong arguments against the creation of yet another form of notation, in the end it proved to be necessary to use a combination of symbols. Efforts have been made to adhere to established standards, notably ISA Standard S5.1 (Instrumentation Symbols and Identification) and ISA Standard S5.3 (Graphical Symbols for Distributed Control/Shared Display Instrumentation, Logic and Computer Systems), but certain exceptions have proved necessary for the very special purposes of this book.

The symbolic notation used is explained fully in Appendix 2.

1.6 UNITS OF MEASUREMENT

The units of measurement used in this book are metric, but because the power-plant industry also spans individuals and countries who have as yet not fully converted to metric units, full conversion factor tables are included in Appendix 1.

CHAPTER
TWO

POWER-STATION PLANT

The history of public-utility electrical power goes back a surprisingly long way. Hard as it may be to believe, the fact is that power stations were delivering electricity to the public in London and New York barely two decades after the end of the American Civil War, and it was within only another 10 years that fairly sophisticated generation and distribution systems employing three-phase generators were operating in Germany.

Although some things change, others stay the same. The generators supplying these early installations were driven by reciprocating engines, but the prime source of power then was steam—exactly as it is today in the vast majority of power stations.

This chapter provides an overview of power-plant basics and of steam characteristics and production. It is merely an overview, provided in an attempt to bridge the gulf separating the discipline of *electronics* from the field of *thermodynamics*, and therefore the approach taken is rather simplistic: it is certainly *not* intended to be an exhaustive treatise on boiler design. It should, however, enable electronics and computer specialists to grasp the essentials of the subject—or at least enough of it to be able to understand the design requirements for boiler control systems.

2.1 BOILER/TURBINE UNITS AND THE RANGE SYSTEM

The power plant as a whole usually consists of one or more 'units', each comprising one boiler whose steam output feeds one turbine, driving a generator. Each unit is a complete entity in its own right, operating separately from the others and only connected together in the sense that the electrical outputs of the generators are connected in parallel to feed the grid. In most cases, there is little or no sharing of common plant, even auxiliaries such as cooling-water pumps or power supply systems. This arrangement yields a high level of security, since failure of an item on one unit will not affect the other units.

In some industrial installations and in older power plant, however, the boilers and turbines are linked via a common steam main. This 'range' system enables some flexibility of operation since any boiler can feed any turbine, and faults in one boiler do not shut down an associated turbine. However, with the introduction of steam reheat, the complexities of trying to apportion reheat steam flow between the operating boilers made the range system impracticable.

Whatever the arrangement, however, the type of power plant discussed here depends on steam for its operation.

2.2 SOME CHARACTERISTICS OF STEAM

There are many reasons why it is necessary to understand the nature and qualities of steam, and how it is produced and used, before any attempt is made to understand how it is controlled—and this is where the electronics or computer engineer must venture into the unfamiliar territories of thermodynamics and mechanical engineering.

Steam, a vapour derived from water—a cheap and plentiful resource—is used in the power generation process because it can be regulated and distributed and because it has good thermodynamic properties. But, in spite of its familiarity and usefulness, steam is quite a complex entity. (Anybody who questions this statement is advised to look at the size of a book of steam tables!)

When cold water is heated, its temperature initially increases in a way that can be detected by human senses (not surprisingly, the heat added in this way is called *sensible heat*). At a certain temperature (100°C at normal atmospheric pressure) the water boils, and from this point adding more heat to the water causes no further increase in temperature: instead, the added energy is used to convert the water to steam. Because heat added after the boiling point has been reached causes no change in the temperature of the water, it is called *latent heat*.

The temperature at which water boils is called the *saturation temperature*, and its value depends on its pressure: the higher the pressure, the higher the temperature at which boiling occurs. (In effect, whatever is above the surface of the water acts as a blanket—pressing down on the surface of the water and holding back the molecules until they eventually gain enough energy to break free at a higher temperature. The higher the pressure, the greater the blanketing effect, and the higher the temperature that must be reached before the water boils.)

The steam produced at the boiling surface is said to be *saturated*, and

8 BOILER CONTROL SYSTEMS

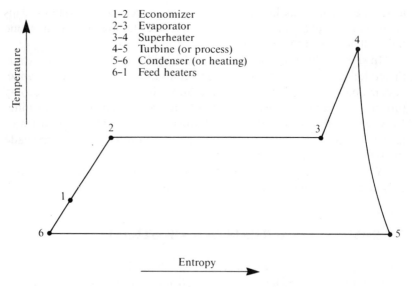

Figure 2.1 Temperature/entropy diagram for the Rankine cycle.

it has the same temperature as the water from which it is obtained. It may contain water droplets (in which case it is *wet* saturated) or it may not (in which case it is *dry* saturated). Once all the water has been converted to vapour, the addition of further heat to the steam increases its temperature—and such steam is said to be *superheated*.

The heat energy of steam is converted (in a Rankine cycle see Fig. 2.1) to mechanical work in a turbine through the process of expansion, and the work that is done is proportional to the pressure range through which the steam expands. It therefore follows that the higher the steam pressure at the beginning of the conversion process (and the lower the pressure at its end), the more work will be done.

In the course of expanding, the steam loses its heat. In a way that is the exact converse of the boiling process, a reduction in the temperature of the steam below the saturation point causes some of it to condense back into water. So, if the steam was initially saturated, the process of converting its heat energy into mechanical work would result in the steam condensing back to water.

This procedure is inevitable—and is an essential part of the closed-loop cycle of the power plant—but it is important that the condensation process should take place only in the area of the plant designed for it (that is, the condenser), since the conversion to water does not occur simultaneously throughout the volume of steam, and hence water droplets begin to appear at random in the steam. If these droplets impinge on hot metal elsewhere than in the condenser, severe thermal shock will occur, resulting in considerable damage. Most importantly, the steam should still be above saturation point when it is imparting the last of its energy to the turbine—which implies that in boiler plant the initial steam temperature must be as high as possible. In other words, it must be superheated.

2.3 AN OUTLINE OF BOILERS

In thermal power plants, the steam used is produced by 'water-tube' boilers, where the walls of the combustion chamber are lined with tubes through which water is passed. (This is unlike the 'fire-tube' boilers used in smaller plants, and once upon a time in steam locomotives, where the hot gases from the combustion processes are passed through tubes immersed in water.)

In these boilers fuel is mixed with the correct amount of air and burned in a combustion chamber. When the amount of air available for combustion is exactly correct for supporting that process—with no surplus or deficit—the proportion of air to fuel is called the *stoichiometric* air/fuel ratio.

The heat released by burning the fuel is used to evaporate water, with the resulting steam passing to the turbine, which drives the electrical generator. The concept is illustrated in a highly simplified form in Fig. 2.2, although this by no means represents how a real boiler functions (if anything, the concept shown is of a 'once-through' boiler).

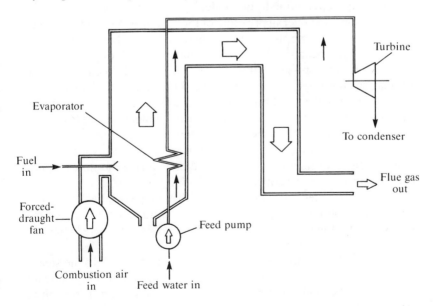

Figure 2.2 Schematic diagram of a simple steam generator.

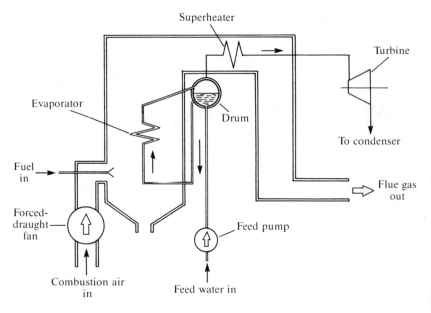

Figure 2.3 Simplified schematic diagram of boiler with superheat.

After the turbine has extracted most of the heat energy from the steam, the vapour is passed to a condenser, where it is finally condensed to water, for recirculation to the boiler.

This is only the overall principle of the boiler/turbine unit. If nothing more were to be done, the energy conversion in such a plant would be very inefficient because the steam would be at a comparatively low temperature and pressure, and, as we have already seen, in practical installations it is important that these qualities should both be high.

A more practical realization is shown in Fig. 2.3. Here the steam is separated from the water in a drum. A 'drum' boiler is one of two general types of configuration: in the other (a 'once-through' boiler), water is converted to steam along a single circuit, roughly as shown in Fig. 2.2, with a small separating vessel being used only during the start-up phase. This arrangement is described more fully later on.

The steam collecting at the top of the drum is passed back to banks of tubes in the combustion chamber, where additional heat is added to it, 'superheating' the steam. The tubes fulfilling this purpose are therefore referred to as the superheater.

Figure 2.4 takes the development a stage further. Here the steam is fed to one stage of the turbine—the high-pressure (HP) stage—and then returned to the boiler for reheating before being used in the turbine again, in this way gaining the maximum benefit from the medium. On

POWER-STATION PLANT 11

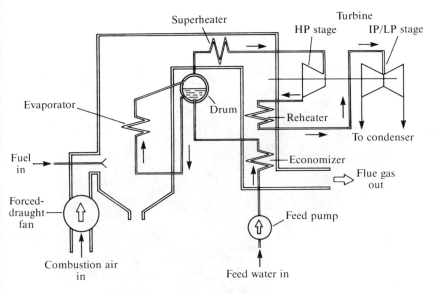

Figure 2.4 Simplified schematic diagram of boiler with economizer, reheater and superheater.

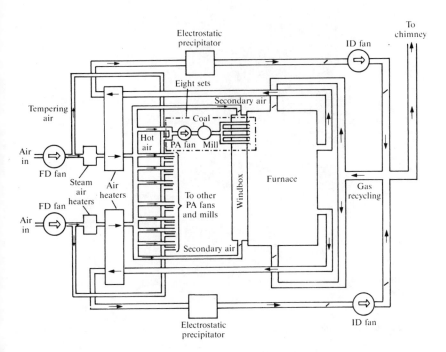

Figure 2.5 Typical air and gas passes of a coal-fired boiler.

12 BOILER CONTROL SYSTEMS

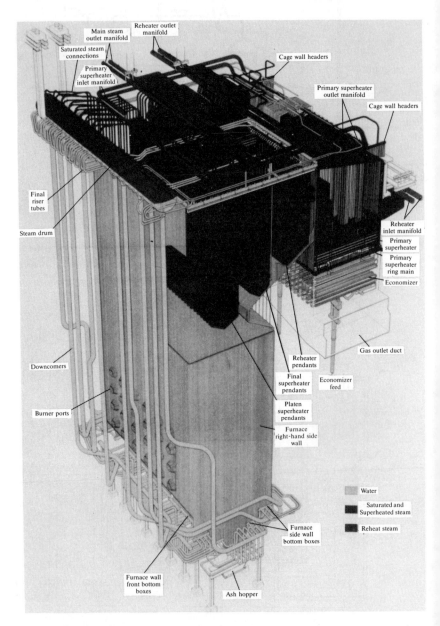

Figure 2.6 Arrangement of a large coal-fired drum boiler (670 MW). (Courtesy of China Light and Power.)

emerging from the reheater, the steam enters the intermediate-pressure (IP) stage and finally the low-pressure (LP) stage of the turbine.

This diagram shows another technique for gaining greater heat recovery from the heat released by burning the fuel: additional banks of tubes prewarming the water before it passes to the evaporative section of the boiler and cooling the exit gases before they are released to the atmosphere. This stage improves the efficiency of the boiler, and is called the economizer.

Of course, all this is only schematic and highly simplified: in reality, plant is considerably more complex than this, as can be seen by comparing this arrangement with the gas and air circuits of a real power-plant boiler (a typical example of which is shown schematically in Fig. 2.5 and in 'exploded view' form in Fig. 2.6). But the principles described above are sufficient for understanding the fundamentals of the plant. They should, in any event, enable the reader to understand the functions of the control loops.

In a subsequent section we will examine the combustion process that produces the heat for warming and evaporating the water, and for superheating and reheating the steam, but first we will look at the arrangements for obtaining the source water and for recovering condensate.

2.3.1 Feed-Water Systems

The water/steam part of the boiler is essentially part of a closed loop, which includes the turbine, the condenser and the auxiliaries. Water is boiled to produce steam, which, having imparted its energy to the turbine, condenses back into water, which is fed back into the system. Losses are made up by adding make-up water. (See Fig. 2.7, which shows the steam and water circuits of a typical boiler/turbine unit.)

However, as stated earlier in this chapter, the objective is to convert the maximum possible amount of heat energy from the combustion process to rotational energy in the turbine, and, to do this, part of the circuit must be at a high pressure. The steam pressure is generated by the heat released in the combustion process, and feed-water pumps force water into the system to replace the high-pressure steam delivered to the turbine.

A power plant generating tens or hundreds of megawatts may operate at a pressure of 16 MPa (say, 3000 pounds per square inch) and use steam temperatures of around 550 °C (1050 °F). The quantities of fluid pumped around such systems are, of course, related to the power output. The actual ratios vary quite considerably according to the design of the plant, the fuels burned and so on, but to provide some indication of the relationships in a large modern plant, a 660 MW boiler requires feed

14 BOILER CONTROL SYSTEMS

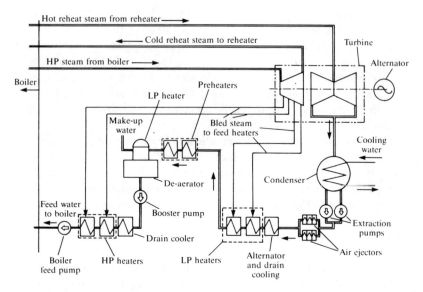

Figure 2.7 Feed-water, condensate and steam circuits of a large boiler/turbine unit (schematic).

pumps delivering a total of over 1100 tonnes (say, 3 million pounds) of water per hour. Pumping water of this amount to pressure of this magnitude demands considerable power: on such a boiler, it would be common to have three feed pumps, each rated to carry 50 per cent of the maximum flow into the boiler—the boiler's maximum continuous rating (MCR). There would normally be three such pumps, two running and one spare, and each would be driven by a motor consuming some 10 MW.

2.3.2 The Steam Circuits

As already explained, in a drum-type boiler the steam leaving the drum is passed back to the tubes lining the furnace walls for the addition of heat, which causes it to become superheated. Depending on the type of boiler, this process may occur in a variety of ways, but in the example shown in Fig. 2.6 the hot combustion gases pass over several banks of tubes. These include a primary (or 'radiant') superheater, where they pick up heat by a process of radiation and convection, and a secondary (pendant) superheater, where further heat is transferred by convection alone.

The superheated steam generated in this way needs to be controlled, because the turbine requires steam of a constant temperature. (In practice, the design of each boiler produces very definite temperature/load characteristics, a matter that is discussed more fully in Sec. 3.8.) Control is achieved by heating the steam more than is actually needed,

(a) Single-stage attemperation

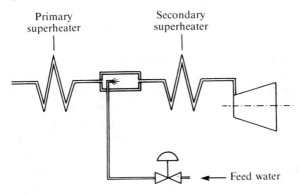

(b) Two-stage attemperation

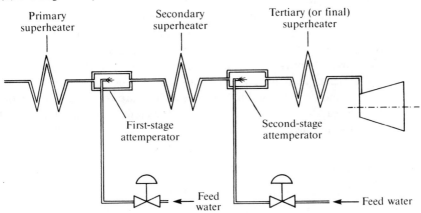

Figure 2.8 Superheater arrangements, with spray attemperators.

and then cooling it in a controlled way. This is done in attemperators (Fig. 2.8), which may be heat exchangers or vessels where (comparatively) cool water is used to reduce the temperature of the steam by direct (or, very rarely, indirect) contact. (It should be stated here that the 'natural' characteristic of the boiler is for the steam temperature at low loads to be low, so that attemperation is only feasible at the higher loads.)

In most cases, the attemperating water supply is obtained from the feed-water pumps.

Attemperation is sometimes achieved in a single stage inserted between two sections of the superheater (as illustrated schematically in Fig. 2.8a), but in some instances there may be two or more stages of attemperation, in which case each stage is inserted between superheating sections (Fig. 2.8b).

The steam emerging from the final superheater feeds the turbine, whose governor adjusts the rate of admission to keep the machine spinning at the correct speed or delivering the required power.

The speed/power relationship of the machine depends on the electrical connections of the alternator that is being driven by the turbine. An alternator that is not connected electrically in parallel with other generators can spin at any speed up to its safe maximum, and the frequency of the resulting alternating voltage is a function of that speed. However, once it has been synchronized (that is, electrically connected in parallel with other generators), the machine spins at a speed dependent on the frequency of those generators and on the number of poles in the alternator. Under these conditions, admission of more or less steam to the turbine controls the power delivered to the turbine, and hence the electrical power dispatched to the system.

Steam re-entering the boiler for reheat purposes is passed to further banks of tubes, again usually with an inter-stage attemperator (although in many cases the reheat temperature is controlled by varying the relative amount of combustion gases passing over the tube banks).

The use of spray water at this stage has an adverse effect on the overall efficiency of the cycle, and a better method of control acts by regulating the flow of combustion gases which are recycled through the boiler by additional fans. These 'gas recycling' fans boost the heat transfer to the reheater section.

2.3.3 The Combustion Process

As stated in the previous section, the fuel is burned in conjunction with air. The combustion process is one in which heat is released through the chemical reaction between the combustible material and oxygen. In thermal power plant, the combustible material is usually a compound of carbon and hydrogen (such as coal, gas or oil), while the oxygen is provided in a supply of air.

Now for a little bit of chemistry! The chemical equations for the various combustion reactions are quite important: they show not only the processes involved, but also how the products of combustion are generated (and these, of course, are the subject of much environmental debate). In addition, they let us evaluate the amount of air that is required for the process. The equations are as follows:

$$C + O_2 = CO_2$$ (the combination of carbon with oxygen to form carbon dioxide)

$2H_2 + O_2 = 2H_2O$ (hydrogen and oxygen combining to form water)

$S + O_2 = SO_2$ (sulphur and oxygen forming sulphur dioxide)

Chemistry textbooks show how to determine the amount of oxygen required for each of these reactions in relation to their atomic weights. The amount of oxygen needed for complete combustion of a fuel is obtained by totalling the individual oxygen quantities, and subtracting any oxygen contained in the fuel itself.

Let us consider, for example, a coal having the following composition (by weight):

Carbon	74%
Hydrogen	5%
Oxygen	5%
Nitrogen	1%
Sulphur	1%
Moisture	9%
Ash	5%

The textbooks tell us that roughly 2.3 kg of oxygen is required to burn 1 kg of this fuel. Given that air contains around 23 per cent oxygen by weight, it is clear that burning this particular coal requires the provision of 10 kg of air.

2.3.4 The Waste Products of the Combustion Process

The above chemical equations show that, whatever the actual quantities involved, the burning of a hydrocarbon fuel yields products of combustion. Typically, these include carbon monoxide, carbon dioxide, sulphur dioxide and water, all of which are carried through to the flue gases leaving the boiler via the chimney.

Since the combustion is effected with air, these products are accompanied by the inert gas nitrogen, which makes up around 77 per cent of air, and by any oxygen in the original air supply that is surplus to that needed for the combustion process.

The oxides of nitrogen are:

NO_2 nitrogen dioxide
NO nitric oxide
N_2O nitrous oxide

The first two are produced in the combustion process. (The last is the so-called 'laughing gas' used by dentists.) As a group, these gases are commonly referred to as 'nitrogen oxides', or NO_x for short, and are usually included in any environmental debates on the effects of power-plant emissions. The facts of life in the combustion process are that the higher the combustion efficiency the higher will be the production of nitrogen oxides: the best and most efficient burners produce the highest nitrogen oxide levels. Minimizing NO_x formation is an art that involves the dynamics and the design of the burner so as to yield adequate combustion efficiency with the production of no smoke and no carbon monoxide.

In addition, most fuels contain incombustible solids, and these too are carried over to the flue gases. The presence of these compounds in the flue gases causes many varieties of problems: if the gases are sufficiently cooled when they come into contact with the ductwork of the plant, they condense into acids, which corrode the metals of the construction; if they are released into the atmosphere, they eventually condense to contribute to acid rain. Furthermore, we now know that the products of combustion can combine with atmospheric compounds in ways that may possibly contribute to global climatic effects.

Our usage of energy expands considerably with the degree of technical and commercial advancement, and when the global population explosion is combined with the thirst of developing cultures, the demand for energy grows enormously. Economists relate energy production and use to a unit that corresponds to the burning of 46 500 million tons of coal. This is the 'Q', and it is approximately equal to 3×10^{14} kilowatt-hours of electrical energy. The world consumption of energy over the period of 1800 years after the birth of Christ totalled probably no more than 4 Q, but by 1850 the rate of consumption had already reached around a quarter of this amount per century, while over the next hundred years it increased to 10 Q per century.

Since, no matter how efficient the plant, there must always be some waste from the power generation process, trivial losses from individual installations are multiplied by the numbers and sizes of plants to cause significant climatic effects.

Clearly, to minimize these effects, the power generation process must be operated efficiently and cleanly. Continual research is leading to the development of new combustion methods (such as fluidized-bed boilers) and to ways of cleaning up the exhaust gases of conventional processes. But in all cases—as we shall see later on—the nature of the plant is such that complex automation is needed in all the processes, to achieve the highest possible level of environmentally benign operation.

Apart from the environmental effects, it is important to recognize the effects on the plant itself of the various chemical compounds involved in the combustion process. For example, a shortage of oxygen, low air temperatures or inadequate mixing of fuel and air at the burners will all contribute to poor flame conditions, where the failure to oxidize the carbon fuels will leave unburned carbon, which will appear as smoke, while other compounds will produce severe long-term corrosion of the tube walls.

2.3.5 Draught Plant

The admission of air into the combustion chamber of a boiler requires the provision of fans to deliver the air to the furnace and (in the case of balanced-draught boilers, which are a feature of most large power plants) to extract the hot gases and convey them to the chimney for discharge to the atmosphere. The fans producing the air for combustion are called *forced-draught* (FD) fans, while those transferring the hot flue gases to the chimney are called *induced-draught* (ID) fans. The control systems for these fans ensure that the flows through inlet and outlet are kept in synchronism, and are usually arranged to maintain a slight suction within the furnace.

In considering the control systems it is important to bear in mind the sheer physical size of the relevant plant. Figures 2.9, 2.10 and 2.11 show

Figure 2.9 Shop-floor assembly of a low-NO_x burner for a 660 MW coal-fired boiler. (Courtesy of Babcock Energy Ltd.)

20 BOILER CONTROL SYSTEMS

Figure 2.10 Inside the combustion chamber of a 570 MW boiler. (Courtesy of Burns and McDonnel Engineering Co., Kansas City.)

the scale of typical burners, combustion chambers and fans of large boilers. To put these items into context, it must be understood that the fans of a 660 MW boiler may be driven by electric motors delivering some 20 000 horsepower and consuming 15 MW of power.

The flow of air through the fans may be regulated by dampers in the inlet or outlet ducts, by vanes at the fan inlets or by varying the speed of the fans.

Air heaters Even after everything possible has been done to transfer the heat of combustion to the incoming water, the flue gases leaving the economizer still contain a substantial amount of heat, and in most

POWER-STATION PLANT 21

Figure 2.11 An induced-draught fan being installed in a 660 MW lignite-fired power station. (Courtesy of Howden Sirocco Ltd.)

installations this is transferred to the incoming air in an air heater. In boilers burning coal, heavy fuel oil or other solids this achieves a valuable secondary feature, since it helps to remove surface moisture from the fuel and thus assists in achieving rapid combustion. But in installations where the fuel is gas or light oil the purpose of the heat transfer is merely to cool the flue gases. In such cases it would be more correct to refer to the air heaters as gas coolers—but custom and practice have long since cast a mould that would now be difficult to break.

The air heaters may be either regenerative or recuperative: the first type uses an intermediate medium to transfer heat from the flue gases; the other uses a direct transfer of heat across a dividing partition. One variety of regenerative air heater is the Ljungström type, where metal plates are first heated by the hot flue gases before being rotated through the inlet air steam.

An important consideration is that the mechanism that results in cooling of the boiler exhaust gas should not be excessive—cooling the flue gases below their dew point would cause acids to condense on the metal at the boiler exit passes.

2.3.6 Fuel Systems

The other half of the combustion process is, of course, concerned with the fuel. In this book we deal with the fuel system *after* the draught plant, for two reasons:

1. To put the two halves of the combustion process into context (the impression is often gained that the fuel system is all-important: by now the reader should have understood the significance of air in the process).
2. Because of the complexities arising through the wide variations in the types of plant (governed by the fuel being burned).

The second point will be apparent in the next four subsections, where the various fuel arrangements are outlined.

Gas and oil firing Liquid or gaseous fuels present fewest problems, since they can be delivered to the burners by comparatively simple means. Even then, however, precautions must be taken to ensure safe and efficient operation. Gas is hazardous, so that care needs to be taken to prevent leakages, and any equipment that might come into contact with the fuel must be so designed and constructed that it can introduce no possibility of accidental ignition.

The oil fuels encountered in power-plant fuel systems range from light paraffin types to heavy, viscous varieties, reminiscent of tar. Both types must be pumped to deliver the fuel to the burners at a satisfactory pressure, but the heavy types must also be heated and continuously circulated to prevent solidification. An important adjunct of such boilers is therefore the fuel-oil pumping and heating set.

Coal firing Solid fuels require more complex arrangements. Although the varieties of solid fuels being burned are very large—ranging from wood to the dried husks of sugar beet (bagasse, known in Australia as megass)—in broad terms here we will be concentrating on coal.

Most coal-burning boilers are designed to handle pulverized fuel—where the fist-sized lumps of the raw material are ground down to an extremely fine powder, which can be transported to the burners in a stream of air.

The grinding process is achieved in mills (pulverizers), which consist of rotating balls or rollers under or between which the coal is crushed (Figs 2.12 and 2.13). The air used for transporting coal through and from

POWER-STATION PLANT 23

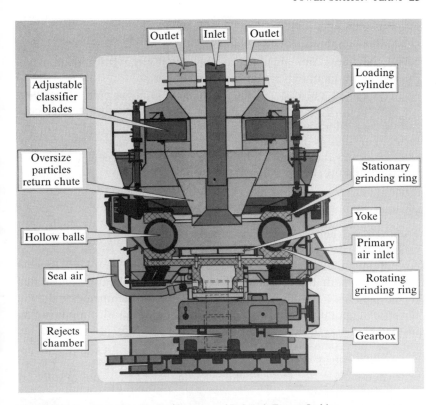

Figure 2.12 Section of a ball mill. (Courtesy of Babcock Energy Ltd.)

Figure 2.13 Pulverized-fuel ball mills in a 660 MW coal-fired power station. (Courtesy of Babcock Energy Ltd.)

these mills is obtained from primary air fans, and it adds to the air supply available for combustion (the remaining air, provided by the FD fans, is called the 'secondary air'). Figure 2.5 showed the gas and air flows through one type of pulverized-fuel-fired boiler.)

The use of pulverizers enables conventional burners to be used for the combustion of the fuel, but raw or only partially screened coal can be used either in chain grates or stokers or in fluidized beds. In the former, the coal is distributed onto a wide moving chain, which slowly rotates to carry the fuel into the furnace, where it is burned. The residue (mostly ash if everything is right) drops off the far edge of the belt into a hopper.

Fluidized-bed boilers In fluidized beds (Fig. 2.14), coal and crushed limestone are fed onto a bed of material that is kept in a state of continuous agitation by the admission of large quantities of compressed air from below. In this condition, the bed material looks and behaves like a fluid, and the coal burning in it does so at a lower temperature than in the equivalent pulverized-fuel burner, so that NO_x production is lower. Moreover, the crushed limestone captures the SO_2 produced in the combustion process.

Fluidized-bed combustion therefore enables a wide variety of untreated fuels to be handled, while yielding fewer emissions of sulphurous or nitrous gases.

Because of the presence of hot, rapidly moving particles which would quickly erode any water tubes exposed to them, a fluidized bed boiler is often constructed with two separate sections: one the combustor itself—with refractory-lined walls—the other made up of one or more heat exchangers where the actual transfer of heat to the water and steam takes place (Fig. 2.15).

Complex fuels Because of world energy resource constraints (and occasional political difficulties), the search for new fuels has gathered pace since the middle of the twentieth century. In many cases, fuels have been evolved from unusual sources or from a combination of established sources. Among the foremost of these currently is 'Orimulsion',[1] a suspension of bitumen in water. Although this fuel requires care in its storage and handling (because of its tendency to become extremely viscous or to decompose, dependent on the temperature), it is a liquid derived from a non-oil resource (bitumen) and it therefore offers a new lease of life to oil-fired power plants designed in the oil boom years of the mid-century.

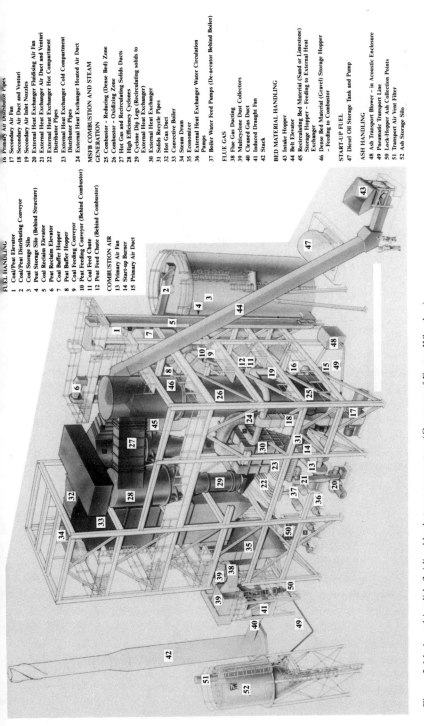

Figure 2.14 A multi-solids fluidized bed steam generator. (Courtesy of Foster Wheeler.)

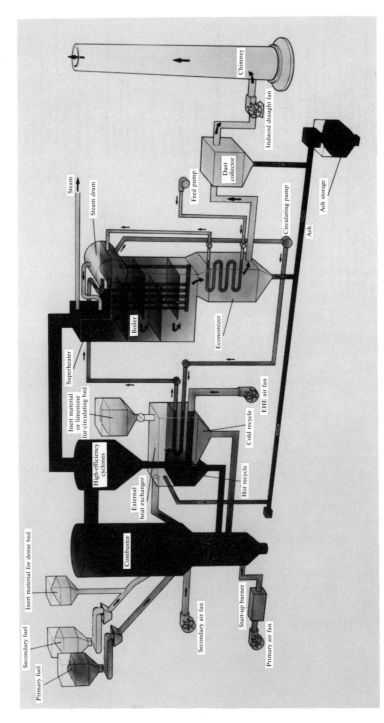

Figure 2.15 Schematic of a multi-solids circulating fluidized-bed boiler. (Courtesy of Foster Wheeler.)

2.3.7 Treatment of Flue Gases

It has already been shown that, whatever the form of combustion, the burning of coal results in the production of flue gases, which necessarily contain significant amounts of undesirable pollutants. Various methods are available for minimizing the effects of these materials.

Filters and precipitators Incombustibles in the original fuel are carried through the furnace and appear as unburnt solids, which, if light enough, get carried into the exhaust gases rather than deposited into bunkers as ash. These contaminants are comparatively heavy, and fairly quickly drop out of the atmospheric exhausts, so any coal-fired plant requires careful treatment of the flue gases if severe pollution of the local environment is to be avoided.

The entrained solids can be separated out by filtration or electrostatic precipitation (or by a combination of both). Precipitators comprise metal rods charged to a high voltage, which attract dust particles in the gases flowing over them. 'Rappers' vibrate these rods to shake off the solids on a regular cyclic basis, the resultant waste material being carried into hoppers for disposal.

Flue-gas desulphurization Filters and precipitators can deal only with solid particles and, for this reason, contaminating gases require other measures if a reasonably clean exhaust is to be assured.

The most severe pollutant is generally considered to be sulphur. Present to some extent in all coals, sulphur can cause damage to the plant and the environment through the processes already described.

The combustion process produces both SO_2 and SO_3, but a lot more of the former than the latter—around 99 and 1 per cent, respectively.

Sulphur can be neutralized by the addition of an alkali such as lime, and a very good way of achieving this is to use limestone fed into a fluidized-bed boiler as already outlined. However, at the time of writing this book, fluidized-bed boilers larger than around 150 MW have not been developed. In any case, it is necessary to deal with the large numbers of conventional coal-fired installations that are already in existence.

Flue-gas desulphurization (FGD) is a process whereby sulphur is chemically removed from the exhaust gases. Although more complex and less efficient than the use of limestone in a fluidized-bed boiler, this process is applicable to the larger plants and can be applied to an existing installation as a retro-fit.

Four types of flue-gas desulphurization process are presently available:

Lime (or limestone) scrubbing	Where SO_2 is combined with lime or limestone to produce gypsum (ultimately used for the production of a building material, plasterboard).
Activated charcoal adsorption	Where the sulphur compounds are removed by passing the gases through a bed of activated charcoal.
Spray drying	Where the gases are combined with a lime solution, and the product evaporated to produce a dry sulphite/fly ash mixture.
Ammonia scrubbing	Where the compounds are combined with ammonia, yielding ammonium sulphate (which is used as a fertilizer).

The first two of these methods result in the production of significant quantities of waste water, which itself requires careful treatment before disposal; in the first the amount of waste water produced is considerable.

Other technologies that have been developed for sulphur removal include the regenerable double-alkali process (which uses a recyclable alkali metal hydroxide) and a combination of sodium scrubbing and thermal stripping.

Nitrogen oxides As already explained, the combustion process produces nitrogen oxides from the nitrogen and oxygen present in the coal and in the combustion air. The quantity formed depends on the temperature of the flame, on the amount of excess air in the process, on the length of time the combustion gases are maintained at a high temperature and on the rate of their cooling. But, whatever the cause, NO_x emissions are argued to be potentially more damaging to the global environment than SO_2 emissions.

Power stations are not the largest contributor to this problem: they produce around 30 per cent of the global emissions whereas motor vehicles generate around 40–50 per cent.[2]

Nevertheless, regulatory and other pressures have forced the power industry to react. Two broad classifications of remedy are available to the industry—modified combustion processes and post-combustion treatment. Figure 2.16 shows one example of a modified combustion process: a burner designed to reduce NO_x emissions by carefully controlling the combustion of the fuel.

Several power plants have been equipped with means of reducing NO_x to nitrogen by combination with ammonia non-catalytically—a process known as selective non-catalytic reduction (SNR)—or within a catalyst bed at the outlet of the economizer—called selective catalytic reduction (SCR). In other installations, the plant has been fitted with flue-gas treatment (FGT) processes.

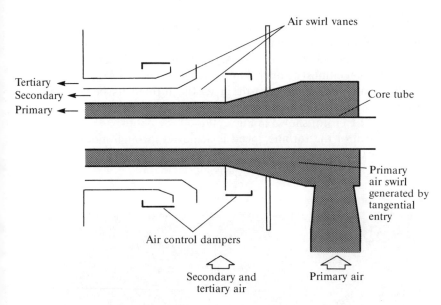

Figure 2.16 Design for low-NO_x burner for a front-wall-fired coal-burning boiler. (Courtesy of Babcock Energy Ltd.)

Fluidized-bed boilers, because of their lower combustion temperatures, produce much smaller volumes of NO_x per tonne of coal than conventional boilers.

2.4 THE STEAM TURBINE AND THE ALTERNATOR

Outside the boiler, the energy of the steam is converted in a turbo-alternator, first to kinetic energy, then to mechanical rotation and finally to electrical energy. Part-way through the process the steam is reheated in the boiler and returned to the turbine.

Once the steam has been expanded through the turbine it is restored to a liquid form, which is then returned to the boiler as feed water. Any losses in the cycle are corrected by the addition of treated make-up water.

2.4.1 The Steam Turbine

In the turbine, the steam is allowed to expand through a series of nozzles, whereby it assumes kinetic energy at the expense of total heat. The jets of steam impinge on aerofoil-shaped blades, imparting rotational motion to the shaft to which the blades are attached. This conversion is achieved partly through the impulse principle and partly through the reaction principle, since the blades—by virtue of their design—act as nozzles.

30 BOILER CONTROL SYSTEMS

Because the steam initially has a high energy content, a high turning moment is imparted to blades near the steam inlet, which are therefore quite short; but at the later stages, when much of the energy has been used, the blades must be made longer to deliver the same turning moment to the shaft. This characteristic gives the turbine its familiar funnel-shaped or waisted appearance, which can be clearly seen in the low-pressure (LP) rotor shown in Fig. 2.17.

Control of the turbine speed is achieved through modulation of a governor valve arrangement.

Figure 2.17 LP rotor of a 350 MW turbine (being installed in the casing). (Courtesy of China Light and Power.)

Steam turbines have to be protected against thermal shock, and against excessive thermally induced differential expansion between the rotor and its casing. The machines therefore have to be run up to speed comparatively slowly—a 660 MW machine may take as much as 24 hours to run up from cold.

2.4.2 Condensate Systems

The steam leaving the turbine passes to a condenser, where a heat exchange with circulating cooling water returns the fluid to its liquid phase. In order to obtain more work from the steam passing through the turbine, the condenser operates at vacuum conditions, the magnitude of which is determined by the temperature of the cooling water and the efficiency of the condenser. The condensate is removed by extraction pumps and eventually returned to the boiler feed system via feed heaters. (A schematic representation of a typical system was shown in Fig. 2.7.)

Modern boilers operate at high pressures, and dissolved oxygen in the feed water will have undesirable effects on the life of the tubes. For this reason, the system has to be carefully designed, constructed and maintained in order to reduce the ingress of air to the absolute minimum level practicable. Part of the design includes the provision of de-aerators to remove any air from the system. Residual dissolved oxygen is absorbed by chemical injection before the water enters the economizer.

2.4.3 The Alternator

The a.c. generator is itself a complex entity. Its output voltage is controlled by adjusting its excitation voltage, while the frequency is determined by the speed of rotation. An alternator connected to an electrical network incorporating other machines must spin at a speed that is a function of the network frequency and the number of its own poles. A two-pole machine operating at 50 Hz rotates at 3000 rpm (50 cycles per second × 60 seconds per minute). At 60 Hz, synchronous speed is 3600 rpm.

The generator—having been run up to speed while disconnected from the supply—is then 'synchronized'. In other words, its frequency and phase is set to match exactly that of the rest of the electrical system. Once synchronized, the machine is connected to the system and begins delivering power according to the steam supply and other conditions.

There are various ways of regulating the power output of a boiler driving a turbo-alternator that is connected to a network, and these are discussed in the next chapter, but the basic objective of all these systems is to regulate the delivery of energy to the turbine.

2.5 ONCE-THROUGH BOILERS

The provision of a drum leads to simplified operation and control of the boiler, since it provides a clear demarcation between the liquid and vapour phases within the boiler circuit. However, a drum is large and expensive, and contains complex internal parts to assist in the efficient separation of the steam from the water (Fig. 2.18). Many plants therefore use the 'once-through' principle, where the evolution of water into steam

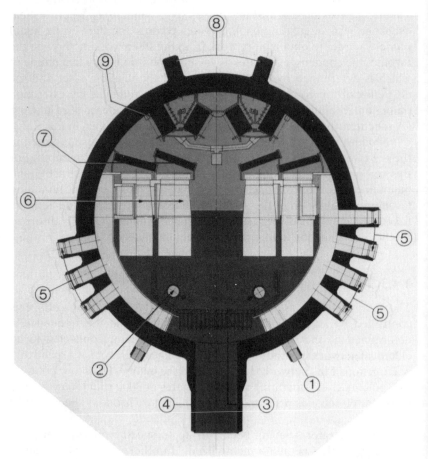

1 Economizer connections
2 Internal feed pipes
3 Debris screen and vortex inhibitor
4 Downcomer
5 Risers
6 Cyclones
7 Primary scrubbers
8 Saturated steam connections
9 Secondary scrubbers

Figure 2.18 Section of a boiler drum showing internal parts. (Courtesy of Babcock Energy Ltd.)

occurs without a drum. Instead, the vessel separating the steam from the water (referred to as a 'separator' or a 'flash tank') is, in general, used only during start-up and shut-down. It is thus small in comparison with a conventional drum, and this leads to an initial cost advantage.

The reasons for choosing a once-through or a drum-type boiler are complicated, but above a certain pressure (somewhat below critical pressure) drum boilers become impracticable owing to the diminishing difference between the densities of steam and water at the saturation temperature.

As the pressure of the steam is increased, the density also increases, and at a certain point its density becomes equal to that of water at the same temperature and pressure. This is known as the *critical pressure*, and at this point the liquid and vapour become indistinguishable. Certain boilers are designed to operate at pressures equal to or above criticality, and these are known as supercritical once-through boilers.

In all once-through boilers, the water/steam interface occurs somewhere along the circuit between feed pumps and the final superheater, and the final steam temperature is a function of the water flow into the boiler (inclusive of any injection at spray attemperators) and the heat input from the combustion process.

2.6 FEED-WATER CONDITIONING

The supply of feed water to a boiler must be chemically treated to remove impurities, and this function must be carried out to exacting standards for the following reasons.

In the course of time, any water-borne solids will be gradually deposited in the evaporative section of a boiler. This is true regardless of the plant type and whether it is shut down or running. As accumulated solids are precipitated, they can cause localized overheating because of their thermal resistance, and this can lead to premature failure of the tubes.

Smaller and older boilers employ 'blowdown' which helps to remove the solids, but it is rarely practicable to follow this procedure on today's boilers—particularly the larger ones. This leads to the need for high standards of water conditioning.

Furthermore, once-through boilers require a precleaning operation with demineralized water before they are started up.

2.7 TURBINE BYPASS VALVES

Turbine bypass valves provide alternative routes for the steam leaving the boiler, avoiding one or more stages of the turbine. As the name indicates,

HP bypasses convey the steam past the high-pressure stage of the machine, while combined HP and LP bypasses route the steam past the whole machine to the condenser.

A bypass is essential for once-through boilers, but it is also very valuable for drum-type boilers, since it helps the operator to match steam and turbine metal temperatures, thereby enabling the turbine to be restarted more quickly. A bypass also enables the unit to recover from turbine trips, since it enables the boiler to be kept steaming while a trip is investigated. If the problem was transient or spurious, the unit can be brought back into service more quickly than if the boiler had also been tripped.

Yet another advantage of a bypass is seen during the initial plant construction programme and after any major refit operation: it enables the boiler to be commissioned in advance of the turbine. Also, any problems with the boiler can be investigated and the solutions tested without any need for starting the turbine.

The turbine bypass function may be achieved by means of one or more steam take-offs beyond the outlet of the superheater, and the total amount of steam being bypassed may be up to 90 per cent of the boiler MCR. If that quantity of steam is bypassed through one circuit, the bypass valve becomes very large indeed. In any case, bypass systems are expensive—though their capital costs are usually quickly recovered by operational savings.

The bypass arrangements are quite complex, since they must incorporate provision for reducing the temperature and pressure of the incoming steam to levels that can be safely handled by the IP stage and the condenser. The change of state from steam to water also presents special difficulties for the measurement system, as we shall see in Chapter 4.

2.8 CYCLIC LIFE-EXPENDITURE CURVES OF TURBINES

Subjecting the high-temperature components of a boiler or turbine to changes in pressure or temperature will cause stresses. If the cyclic changes are greater than allowed for in the design, they will contribute towards the premature demise of the components.

In practice, the effects of these stresses are potentially more severe in turbines than in boilers. For example, analysis has shown[3] that a high-pressure boiler drum, although some 18 cm thick, can safely be subjected to positive or negative changes in saturation temperature of the steam and water inside it of 225°C per hour for an unlimited number of cycles.

Changes of comparable severity would be very detrimental to the life of a turbine, and manufacturers therefore publish 'cyclic life-expenditure' curves, which relate the life of the machine to the number, rate and extent of temperature cycles to which it is subjected.

These limitations have particular significance when it comes to the control of the unit during start-up and shut-down operations. Boilers incorporate very limited facilities for the control of steam temperature during start-up and low-load operation, and it has therefore been common practice to calculate the stresses imposed on the turbine by changes in temperature and to limit them by applying enforced 'holds' on the rate of load raising.

2.9 COMBINED-CYCLE PLANTS

The efficiency and versatility of the power generation process can be improved by combining two or more conversion processes. A combined-cycle plant is one in which, for example, waste heat from a gas turbine is used in a boiler (perhaps with some auxiliary firing), which itself provides steam for a turbine. In such a case, electricity is provided by the generators driven by the steam turbine and the gas turbine. Such an arrangement is shown in highly schematic form in Fig. 2.19.

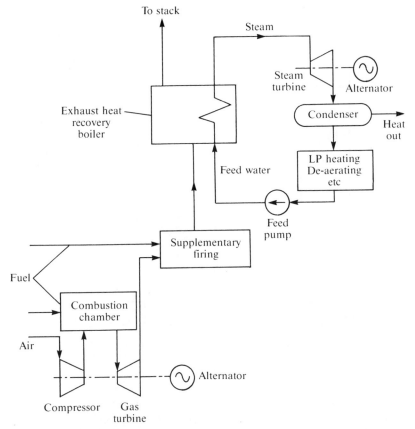

Figure 2.19 Combined-cycle flow diagram (simplified). (Courtesy of Bailey Controls Co.)

36 BOILER CONTROL SYSTEMS

Compared with single-cycle plants, combined-cycle operation offers a higher overall thermal efficiency and lower thermal pollution (less waste heat discharged to the atmosphere, rivers or oceans). It also offers a wide range of flexibility of power generation, and quicker start-up and greater responsiveness to load changes.

2.10 CO-GENERATION

Any thermal power plant must generate some waste heat. In the past this was discharged either to the atmosphere through cooling towers or to

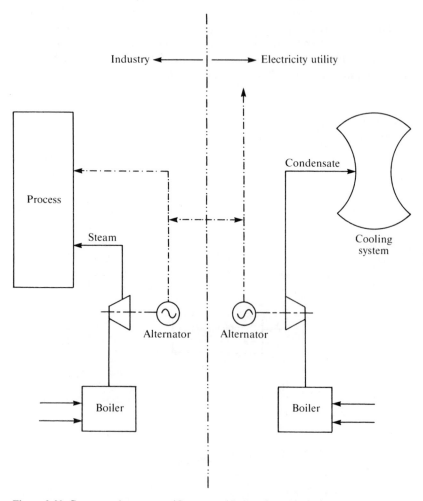

Figure 2.20 Co-generation system. (Courtesy of Bailey Controls Co.)

nearby oceans or rivers, but with the advent of high fuel costs (and with a growing public awareness of the local and global effects of these discharges) the economic balance has shifted in favour of making the maximum possible use of waste heat.

Where the plant is near a user of heat (whether actual or potential) it is possible, and economically desirable, to sell the surplus energy in the form of hot water or steam—to the advantage of the plant owner, the consumers and the environment. The user of the surplus energy could be a large industrial complex or a housing development, or perhaps a large-scale group of agricultural greenhouses. In many cases, however, the recipient of the heat energy is an industrial process that is closely allied to the owner/operator of the power plant.

This process is known as co-generation, and it involves the simultaneous production of electric power and *useful* thermal energy from one or more heat sources. A typical example is shown in Fig. 2.20.

Not a new concept, co-generation lay fallow for several decades and to a large extent its prospects improved only with the post-1970s rise in fuel prices. The appearance of environmental issues on the agenda of public debate also accelerated the emergence of the technology to the point it has now reached, since co-generation (especially when it is based on combined-cycle operation) yields levels of thermal efficiency that are very much greater than those available with conventional power-plant operations.

NOTES

[1] 'Orimulsion' is a trademark of Bitumenes Orinoco SA, a subsidiary of Petroleos de Venezuela SA.
[2] Source: *Power*, Special report on NO_x emissions, September 1988, McGraw-Hill Publications.
[3] Reported by A. H. Rudd and O. W. Durrant at the *American Power Conference*, April 1974.

CHAPTER
THREE
MODULATING CONTROL SYSTEMS

As stated in Chapter 1, the boiler/turbine unit is a complex *interactive* entity. What this means in practice is that variations in many of the controlled parameters will be accompanied or followed by directly associated disturbances in others.

For example, while an alteration in the opening of a feed valve will have an expected effect on the water level in the drum, it will also result in a greater cooling effect within the boiler's water walls, and (in a once-through boiler) this will eventually have the unexpected effect of altering the steam temperature. Similarly, altering the heat input to the boiler will inevitably affect one section of the water tubes more than the others, and this will cause expansion of the water in that region, which will in turn tend to cause the water level in the relevant region of the drum to surge above the mean level over the whole drum. In addition, in this particular example, the release of extra heat within the combustion chamber will tend to produce a transient increase in the furnace pressure, while some direct effects may also be expected in the steam temperature, which will need to be corrected by the loops controlling that parameter.

With reasonably effective control, most of these results will be transient. After a period of disturbance, the various parameters will settle down to the values they held before the change was introduced. However, in the meantime the changes will have been detected by the relevant control loops which, unless they are suitably interconnected, will begin to take short-term action that may not be necessary in the longer term. In extreme cases these interactions can lead to instability.

This is why boiler control systems should have, in addition to the loops regulating individual parameters, adequate feed-forward and feed-back signals to provide anticipatory control (which tries to predict the effect on one part of the system due to a change in another and takes action to reduce the consequences) or compensation (which adjusts the operational parameters to minimize the disturbances after they have occurred). All the main control loops, however, must respond to a central command structure, which sets their individual demands to match the throughput requirement of the plant.

As far as the boiler itself is concerned, the central determinant for the control systems is the demand for steam. This parameter sets the required flow rates for fuel, combustion air, induced-draught gas and feed water, and is consequently the orchestrator for all the other control loops of the boiler (and, in some examples, for the unit). It is hence referred to as the *master* control signal.

3.1 THE MASTER CONTROLLER

In practice, the method of deriving the master signal varies quite considerably from boiler type to boiler type and between control system vendors. The variety and complexity of these various approaches preclude a detailed analysis of how each and every system operates. Instead, an overview is provided of the basic principles, with some discussion of a few options.

In the following discussion, for reasons of simplicity, it is assumed that the master demand signal remains at a fixed value, which is determined by the operator. In practice, however, the demand must be related to the electrical load on the machine, and a brief analysis of the present situation and foreseeable trends is therefore warranted here.

Although present trends towards the use of renewable energy, combined-cycle or small gas-fired plants may eventually lead to national power systems that include a significantly large number of small-sized units, the picture in the current climate is very different. The need to have the highest possible degree of economic efficiency, coupled with the economies of scale, have in the past enforced a trend towards the use of large power-plant units and for these to have large-capacity auxiliaries. Because of this, the minimum power output that can be produced by any single unit is necessarily restricted, and the ability to vary the power production over a wide range requires a level of automation that has not yet been generally achieved.

For example, the pulverized-fuel (PF) mills of a modern coal-burning boiler have a capacity of 80–100 MW, and with each mill capable of turning down stably to (at best) only 50 per cent of its maximum throughput, the maximum change in power that can be accepted by such a unit *without altering the number of mills in service* is no less than 40 MW.

In other words, while it is possible to alter, over a very wide range, the power produced by such a unit, doing so from a central—and therefore usually remote—load dispatcher (or grid control centre) cannot be implemented without the dispatcher bringing mills into service to meet a rising demand, or taking them out to meet a falling load. Given the complexity and hazards of mill operations (see Sec. 3.3), this is not generally considered to be a viable option (although increased and improved automation of start-up and shut-down operations should eventually contribute to changing this situation).

In the meantime, the norm is one of manual adjustment of the boiler throughput (of course, with reference to target load indications from the central dispatcher). This manual adjustment involves trimming the throughput of the plant within the capacity of the operating auxiliaries, and anticipating the changes that will be needed in the number of those operating auxiliaries in order to meet further changes. This *modus operandi* is assumed in most of the following passages.

3.1.1 'Boiler-Following' Mode of Operation

A common method of operation uses the turbine governor as a fast-acting load controller as shown in Fig. 3.1. Changes in grid frequency and power demand are met by adjustment of the turbine governor, leaving the boiler to control its own pressure. (The symbols used on Fig. 3.1 and subsequent figures are given in Appendix 2.)

With this system, the plant operates with the turbine throttle valves partially closed. A short-term increase in electrical demand is met by opening the valves, in this way drawing on the stored energy in the boiler to meet the changed demand. Conversely, short-term reductions in load are met by closing the turbine valves. These actions initially cause

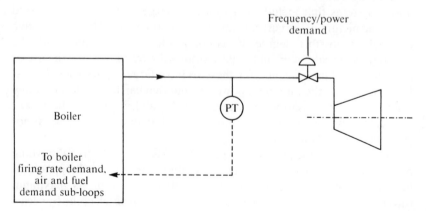

Figure 3.1 'Boiler-following-turbine' mode of master control.

MODULATING CONTROL SYSTEMS 41

alterations in the steam pressure, which are detected, and in turn corrected, by the boiler's master pressure controller. The latter device[1] compares the measured steam pressure with a set value, and compensates for the changing pressure by adjusting its output to call for an appropriate change in the firing rate of the fuel, and for corresponding changes in all the related parameters. (The ways in which these parameters are altered are complex in themselves, and are discussed in depth later.)

With this method the boiler control system endeavours to maintain a fixed steam pressure at all times, and changes in load demand are met by adjusting the turbine inlet valve first, leaving the boiler to respond later. Consequently, this method is called the 'fixed-pressure, boiler-following' (or 'boiler-following-turbine') mode of control.

The system capitalizes on the boiler's stored energy capability—depositing or drawing energy as needed for meeting changes in demand—but it is important to appreciate that it is only small-scale changes in demand that can be met in this way, and that the rate of change of load that can be accommodated is limited by the pressure drop present or allowable across the throttle valve.

The system provides good control of the unit, but it suffers from some limitations:

1. Instability can result under certain conditions (because of the positive-feedback effect of the pressure and flow changes combined with the high gain of the system).
2. The amount of stored energy that is available for meeting changes in demand is limited, and thus the performance deteriorates with increasing periods of mismatch between electrical demand and boiler steam generation.
3. The system contains intrinsic inefficiencies, because of the losses inherent in the throttling of the turbine valves (an action that results in the steam pressure being held at a higher value than is necessary).

The last factor is of some significance for every unit operated in this way; but when a typical grid system is considered, the cumulative effect is to increase capital costs because at any given instant a proportion of installed plant capacity is not being used.

3.1.2 'Turbine-Following' Mode of Operation

An alternative, 'turbine-following-boiler' method sets the boiler's fuel, air and feed-water flow rates at a value relevant to the demand, and uses a pressure control loop, acting on the turbine inlet valves, to maintain a fixed pressure at the first stage of the machine (Fig. 3.2).[2]

42 BOILER CONTROL SYSTEMS

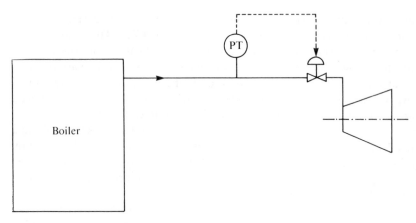

Figure 3.2 'Turbine-following-boiler' mode of master control.

This system is sometimes called 'passive' since it does not respond directly to changes in the alternator load or demands from the scheduling system. Any changes that affect the boiler heat-production rate (such as changing calorific value, moisture or ash content, or feed-water temperature) will affect the steam flow and, ultimately, power generation. For this reason a unit operated in this way will almost certainly contribute to grid frequency disturbances, and a more practicable implementation is to use an 'active' system, where the boiler is controlled to keep the generated power at an operator-set value, as shown in Fig. 3.3.

This arrangement in the long term will maintain generation at the desired value, but the long response time of the boiler means that short-

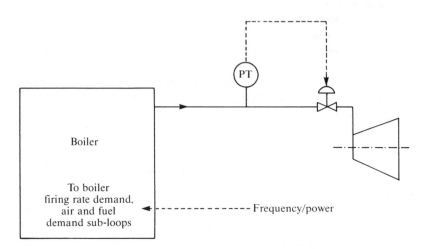

Figure 3.3 Active 'turbine-following-boiler' mode of master control.

term disturbances are fed into the grid in exactly the same way as the passive system described earlier. In addition, since frequency deviations do not affect the load setpoint, units operated in the turbine-following regime cannot be used in a network frequency support mode.

3.1.3 'Sliding-Pressure' Mode of Operation

With the systems described above, the cycle efficiency at part loads is reduced because the valves introduce pressure losses and the feed pumps consume more power than necessary. One way of overcoming this problem with a basic boiler-following system is to set the master pressure, not at a fixed level, but at a value determined by the electrical load. Figure 3.4 compares this characteristic with the fixed-pressure mode of control, while the control system whereby it is implemented is shown schematically in Fig. 3.5.

This configuration results in the boiler operating at a steam pressure that varies according to the load, and it is therefore referred to as 'sliding pressure' or 'variable-pressure' control[3].

It should be understood that sliding-pressure control implies the use of electrically operated 'programmed' safety valves, since simple spring-loaded valves may be unable to respond quickly enough if a sudden, dangerous, increase in pressure occurs while the pressure setpoint is at a low value.

Clearly, sliding-pressure control results in more efficient operation at lower loads since there is minimal throttling-back action by the turbine throttle valves (which in this régime are relegated solely to the role of overspeed protection). However, there are additional benefits associated with variable-pressure operation, two of which are as follows:

1. The unit can be loaded more quickly, since it becomes easier to ensure that boiler and turbine metal temperature rate-of-change limitations are met.
2. The turbine can be started and loaded at lower temperatures and pressures.[3]

On the other hand, *because* of the relegation of the rôle of the throttle valves, variable-pressure operation restricts the ability of the unit to use the stored energy of the boiler to meet short-term changes in demand, and the overall response to short-term load changes is slower than with either boiler-following or turbine-following control.

The concept of the system itself indicates the solution to this problem. Since there is a correlation between the generator output and the turbine first-stage steam pressure, the desired value for the pressure could be derived from the pressure at the turbine HP inlet rather than

44 BOILER CONTROL SYSTEMS

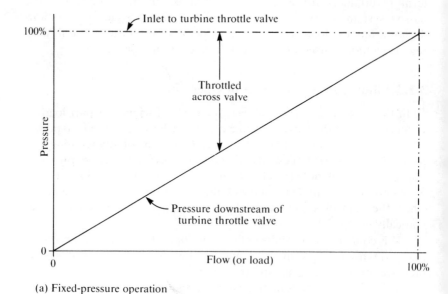

(a) Fixed-pressure operation

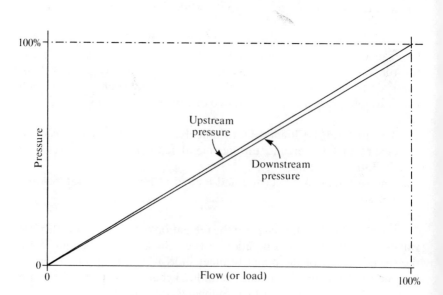

(b) Pure sliding-pressure operation

Figure 3.4 Comparison of fixed- and sliding-pressure operation.

MODULATING CONTROL SYSTEMS   45

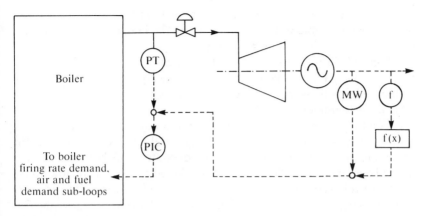

Figure 3.5 Basic 'sliding-pressure' mode of master control.

from the electrical output of the machine. But this is equivalent to keeping a constant differential pressure across the turbine throttle valves, and a simpler way of achieving the same objective is to modulate the boiler to maintain the throttle-valve differential pressure at a fixed setpoint. By maintaining an adequate differential across the valve this system achieves the objective of assisting the unit to respond quickly to load changes.

This 'modified sliding-pressure' mode of operation is shown in Fig. 3.6 and it achieves most of the advantages of the pure sliding-pressure system while retaining much of the responsiveness of fixed-pressure

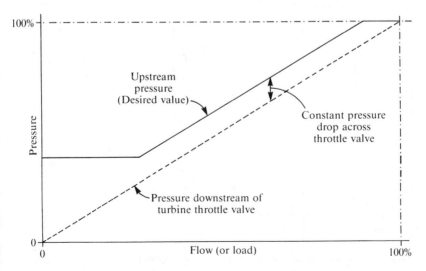

Figure 3.6 Modified sliding-pressure operation.

operation. (The pressure/flow characteristic is flattened out at the lower flows to avoid problems that may arise with the boiler circulation system if the pressure falls to too low a value.)

Figure 3.7 shows how the objective is achieved. The turbine throttle valves provide a fast response to load changes, but the pressure setpoint of the boiler is now related to the unit load demand or the steam flow, whichever is greater.

Load changes are initially met by rapid adjustment of the turbine throttle valves, while changes in unit demand are met by altering the firing rate of the boiler. With this mode of control, the boiler pressure is altered over part of the flow range, but held steady at high and low flows.

With this system, it is necessary to ensure that the boiler pressure setpoint is not adjusted as soon as the unit load demand changes, since (as will be seen later on) the fuel control system is itself load-related, and the combination of these systems could result in pressure overshoot.

In practice, in common with any master pressure control loops that embody a steam-flow component, this concept requires the incorporation of systems to prevent instability due to the positive-feedback effect of the flow signal.

To examine how this effect occurs, consider what happens when the boiler operator introduces an additional bank of burners. The transient increase in steam pressure that results causes an increase in the steam

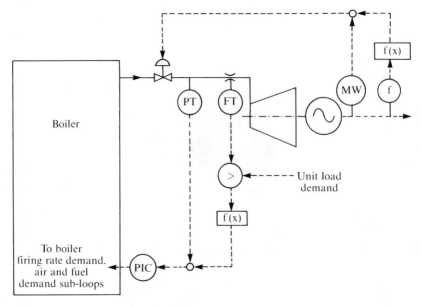

Figure 3.7 Modified sliding-pressure mode of master control.

flow. Uncompensated, this flow would look like an increase in turbine demand and this would in turn demand more firing—when, in fact, *less* firing is what is needed.

In an endeavour to obtain better response, some configurations have included sub-loops to compensate for the changes in the throttle valve/power relationship that result from variations in the state of the turbine auxiliaries. But such complexities make the system more difficult to commission and do not always yield the anticipated benefits.

Even more complex systems have been devised for obtaining optimum performance from the unit as a whole, and these have proliferated with the advent of more powerful computers and the availability of extended and improved measurement techniques. The philosophy behind all of these concepts is to take into account the status, condition and response rates of the critical plant items and their auxiliaries.

Necessarily, these systems couple together the control of both the turbine and the boiler, and for this reason the arrangements are called *coordinated unit master* controllers. Various configurations of such systems have been developed, many for specific types of plant and mostly for once-through units.

One method is based on the boiler-following mode of control, but with limits applied to the turbine load requisition when the boiler's capabilities are unable to keep up with the rate of change or magnitude of the demand. Another uses carefully derived simulations of the boiler dynamic response to feed to the turbine throttle valve a signal that represents the actual capability of the boiler.

These two arrangements are described more fully below.

Selective boiler and turbine following In one system, a turbine-following master is used, combined with a modified sliding-pressure mode of operation. However, this régime is followed only to a partial extent, the turbine throttle valves being set to give a predetermined pressure drop at all loads. The boiler master pressure setpoint signal is obtained by combining the power demand with the output of a controller that compares the pressure drop across the throttle valves with a fixed setpoint.

An increase in power demand is met, therefore, by adjusting the boiler firing-rate demand but, since the boiler response is slow, the throttle valves are opened as well.

The system combines the features of both the boiler-following and turbine-following modes, to give rapid response to small-scale load changes, while minimizing the feed-pump power consumption at part loads. The system also lengthens the life expectancy of the high-temperature surfaces by reducing the pressure stresses.

48 BOILER CONTROL SYSTEMS

Masters employing boiler simulation models More complex master systems have been developed, using small-scale simulations of the system dynamic responses to anticipate behaviour of the plant and to take appropriate action. Various subsets of this technique have been evolved to deal with the different requirements (when sliding pressure operation is optional, whether it is in use; whether the thermal storage capacity of

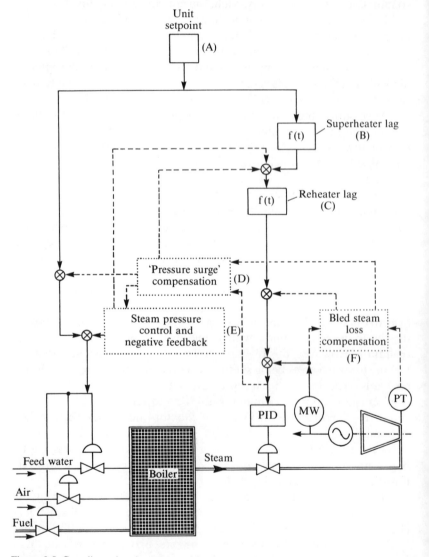

Figure 3.8 Coordinated unit master with plant modelling (no usage of boiler stored energy).

the boiler can be used; the amount of bled steam being drawn off for feed heating; etc.).

One version of this system is shown in Fig. 3.8. This system is used with once-through boilers to overcome the lack of stored energy in such installations. In this system,[4] under steady-state conditions, the unit load setpoint (item A on the diagram) is applied to the boiler and turbine in parallel. For the boiler it forms the setpoints for the fuel, air and feedwater systems, while for the turbine it forms the setpoint for a closed-loop power controller modulating the turbine throttle valves.

If the unit setpoint is changed, the new command is fed directly to the boiler, but via time-delay functions to the turbine (since the boiler has the longer time constant). By using two separate time-function blocks in the turbine loop, the system allows for optimization of the control response. The first of the blocks (B) is set up to correspond to the response time of the evaporative and superheating sections of the boiler, while the second (C) simulates the delay in the reheater. The overall result is that increases in the demand for electrical power from the unit are applied to the turbine only when the boiler is capable of responding to the demand for more steam.

A subsystem (F) compares the pressure at the inlet to the turbine HP stage with the electrical power being generated, and compensates for bled-steam usage by trimming the power controller setpoint. Another subsystem (D) stabilizes the overall response of the system against the 'pressure surge' that occurs when the stored energy within the system is altered to meet sudden changes in the electrical load.

Masters used with plants lacking separating vessels In once-through boiler systems, the master signal has to dictate the operating parameters of the major control loops in a very specific way, and in boilers where there is no vessel for separating the steam from the water these systems attain a special importance. One version (shown in Fig. 3.9) regulates all the major functions of such a boiler, by feeding the unit load demand signal in parallel to all the following subsystems:

- the turbine inlet valve;
- the main steam valve;
- the boiler demand (i.e. firing rate and feed-pump speed);
- the steam-temperature loops;
- the flash-tank level;
- the LP bypass.

In this system, the unit load demand is derived from the automatic load dispatcher, though an auto/manual station enables the signal to be generated manually if desired. Within the block labelled 'unit load

50 BOILER CONTROL SYSTEMS

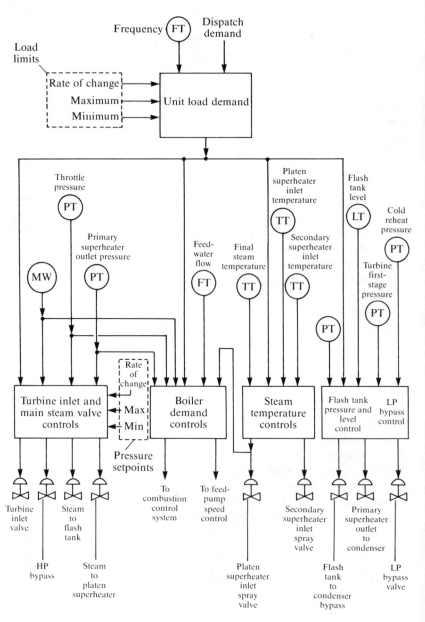

Figure 3.9 Unit control of a plant with a supercritical once-through boiler. (Courtesy of Bailey Controls Co.)

demand' a system is provided that duplicates the action of the turbine governor to prevent the system from countermanding control actions by the latter during frequency disturbances. The system also continuously monitors the boiler and turbine auxiliaries and 'freezes' or runs back the load demand if any operational constraint is reached or exceeded.

The system also incorporates an intelligent loading ramp generator, which holds the demand steady during the boiler transition mode, when the steam flow is entirely through the turbine bypass valve(s), until the flow is completely diverted to the turbine.

With this type of unit the superheater is the only part that can operate under the variable-pressure régime, a function that is achieved by throttling a valve at the platen superheater inlet, reducing the amount of throttling necessary at the turbine inlet valves. This action does not result in any saving in feed-pump power but it does contribute to extending the life of the superheater.

A more detailed discussion of once-through boiler feed-water controls is provided in Sec. 3.7.2 (which also provides a more complete explanation of the start-up and operation of this type of supercritical once-through plant).

3.2 COMBUSTION CONTROL

Although no single aspect of boiler control may be considered to be simple or safe, it is in the area of combustion control that the greatest care must be exercised. This is because mistakes in the burning of the fuel can create the greatest hazards to the plant and its operators.

Surprising as it may seem, this is true irrespective of the fuel being burned. While it may be clear that the burning of gas requires special care and that oil firing can be hazardous, it may not be quite so obvious that the combustion of solid fuels such as coal or the dried husks of sugar-cane is also dangerous. In fact these are also hazardous operations, and so some explanation of the risks is warranted.

At the beginning of this book it was explained how coal was ground down to resemble a fine—but black—talcum powder. When this innocent looking dust is mixed with the primary air supply and blown in great quantities along the large-diameter pipes to the burners, it forms an explosive mixture. A respectably sized PF burner will be capable of handling over 4 tonnes of this material every hour—say 1 kg per second—and there may be 60 such burners in a single boiler; so that around 60 kg of coal is being ignited—releasing a billion joules of energy—in each *second* of operation of such a boiler. In rather unscientific, if dramatic, terms, this is equivalent to the energy released by detonating a thousand sticks of dynamite a second—not an activity for the faint-hearted!

Safely controlling a process releasing this energy is a demanding problem, and it is one whose nature depends on the fuel in question—the combustion control system of a coal-fired boiler being significantly different from that of a gas-fired or an oil-burning plant.

In the following section, the basic requirements are first analysed. The text then outlines how these are met in the combustion control systems of the simplest configuration—an oil-fired plant—and moves on through gas-fired installations to coal firing.

3.2.1 The Fuel/Air Ratio

Before examining the principles of air-flow control, it is worth while considering some of the requirements and practicalities of the combustion process.

When burning any fuel, there is a fixed relationship between the amount of fuel being burned and the amount of air required to support combustion. In practice, because of unavoidable imperfections in the mixing of fuel and air, the use of the theoretically correct ratio of air to fuel (the stoichiometric ratio) will result in incomplete combustion of the fuel—leading to the production of black smoke and poisonous carbon monoxide, and the risk of accumulating a dangerous collection of unburnt fuel in the plant. Therefore, a certain amount of excess air is always necessary, and it has been reported[5] that in practical large multiburner coal-burning furnaces the excess air must be maintained at around 10 per cent at the burners in order to ensure complete combustion with the formation of no more than a trace of carbon monoxide in the furnace.

Although this excess air is necessary, it must clearly be minimized— not only because excess air contributes towards the formation of undesirable emissions from the stack, but also because too much excess air is undesirable since it adversely affects the boiler efficiency by causing unnecessary heat to be wasted via the chimney. Minimizing the amount of excess air also helps to reduce the amount of fuel-borne sulphur that is converted to SO_3, especially in oil-fired boilers.

The total heat losses in a boiler take three forms:

Chimney loss   that is, the heat in the exhaust gases.
Unburned loss   the heat theoretically available in the fuel but which is not converted due to incomplete combustion.
Radiation losses

MODULATING CONTROL SYSTEMS 53

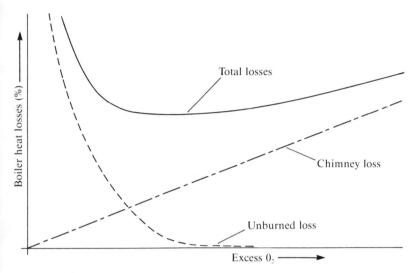

Figure 3.10 Losses related to excess oxygen in furnace.

As shown in Fig. 3.10, the heat losses are affected by the amount of excess air provided. The diagram illustrates the point that losses are high at low air/fuel ratios because of incomplete combustion, while at high ratios they again increase because of the rising power consumption of the fans (which reduces the net electrical efficiency of the unit). From this, it would appear that there is an optimum operating point for the system in order to minimize the heat losses. Unfortunately, the reality is more complicated, and there has been much debate as to the best form of fuel/air ratio 'trim' system. However, one thing is clear: the optimum for minimizing the total losses usually involves the production of carbon monoxide and some smoke. (This is because the chimney temperature is also affected by the excess air level, so that the chimney losses are more sensitive to changes in excess air level.) Therefore, the operational excess air level is not set to achieve optimum heat losses, but to reduce those losses to the minimum possible value before carbon monoxide and smoke are produced.

In addition, if poor design of the plant results in operating with very significant amounts of excess air, the capital cost of the installation is also increased, since the fans are larger than necessary.

Obtaining optimum conditions is also complicated by the uneven distribution of, and the leakages of, air within the boiler as a whole. Figure 3.11 shows the origin of the air within various sections of a medium-sized plant with two pulverizers (which grind the coal before it is burned) and four burners. The diagrams show how the air in the furnace

54 BOILER CONTROL SYSTEMS

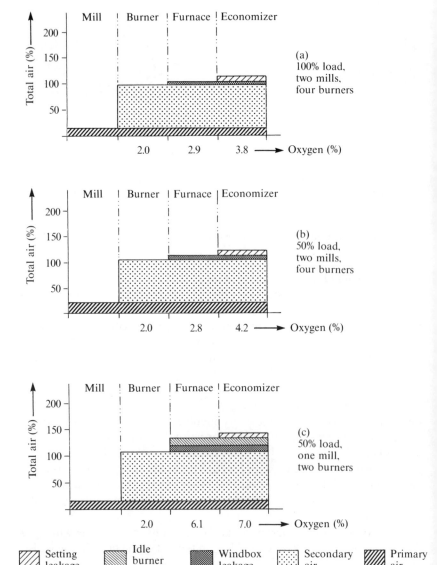

Figure 3.11 Sources of air in furnace. (Courtesy of Burns and McDonnel Engineering Co., Kansas City.)

MODULATING CONTROL SYSTEMS 55

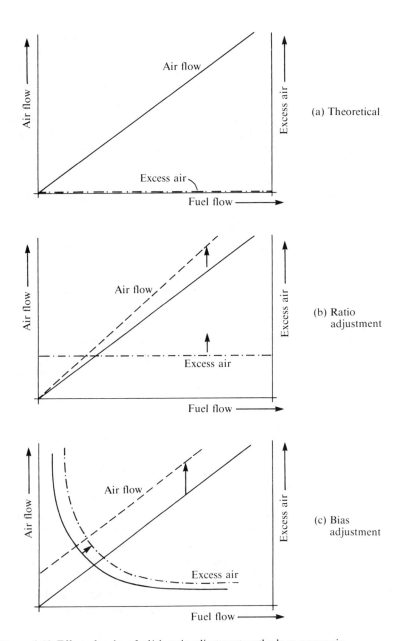

Figure 3.12 Effect of various fuel/air ratio adjustment methods on excess air.

varies with boiler load, and how it is affected by the plant configuration selected to meet the load (that is, either operating the full complement of pulverizers and burners at reduced throughputs, so that each operates at 50 per cent of its capacity, or achieving the same objective by running half the complement flat out). The diagram illustrates the presence of 'tramp air' (or the 'setting leakage'), which is due to the natural ingress of air through observation ports, soot-blower openings and other orifices. It shows that it is undesirable, at least from the point of view of maintaining a low flue-gas oxygen content, to meet part-load conditions with less than the maximum possible number of pulverizers in service since this results in the entry, through idle burners, of air that will not be available to support combustion at the operating burners.

Whatever the target criteria, there are various ways of achieving the desired fuel/air relationship. By adjusting the ratio between the flows, it is possible to achieve an amount of air that exceeds, by a fixed percentage, the amount required for complete combustion—as illustrated in Fig. 3.12a. Plotting the resultant quantity of excess air against boiler load will show a flat characteristic as shown in Fig. 3.12b, and by varying the ratio setting (for example, to the broken line) the excess-air line is moved vertically.

In practice, this characteristic is difficult to achieve, since leakage through dampers and registers will be proportionally greater at low loads, causing the excess air *vs* load graph to curve upwards at the lower loads.

Alternatively, adding a fixed bias to the signal demanding air flow (Fig. 3.12c) causes the excess air *vs* load graph to assume a hyperbolic characteristic (Fig. 3.12c) since the fixed bias represents an infinite oversupply of air at zero oil throughput, and its effect diminishes as the fuel flow increases. In this configuration, varying the bias moves the fuel/air ratio bodily upwards, with the effect of moving the focus of the hyperbolic excess-air curve further away from the intersection of the axes (as indicated in the figure).

Gas analysis It is nowadays common practice to operate not with a fixed fuel/air ratio but with one that is adjusted automatically in response to some form of on-line chemical analysis. The objective is to maximize combustion efficiency while at the same time minimizing undesirable flue-gas emissions, and to do this over a wide variety of plant operating conditions. However, there is some argument over which is the best technique to use. Examples include monitoring of the following:

- oxygen;
- carbon monoxide;
- opacity;
- carbon in ash.

Oxygen trim As indicated earlier, the excess oxygen in the flue gases provides a reasonable indication of the point where the maximum efficiency of operation is obtained, and this is logical, since complete combustion results in the total combination of all oxygen with the fuel.

This principle is used in the basic oxygen-trim system, wherein a signal from a flue-gas oxygen analyser is compared with a set-value signal (which may or may not be load-related) and the error fed to a three-term controller, which adjusts the fuel/air ratio until the error is eliminated. (It should be noted that, in practice, the larger boilers require the use of several analysers in order to gain a reasonable average reading across their large flue-gas ducts.)

In many instances, the O_2 setpoint is adjusted in relation to the steam flow, since the optimum oxygen content of the flue gases varies over the boiler load range.

Oxygen trimming works adequately in many installations, but suffers from the disadvantage that poor performance of one burner (leading to a high level of excess air at that burner) will cause the oxygen controller to reduce the air to all burners equally, even though the others were previously operating correctly. However, in cases where multiple analysers are provided, it may be possible to relate a measurement to a particular burner or group of burners in a multiburner furnace, and by this means to detect and correct mal-operation of an individual burner.

Another problem is the errors that are introduced through the presence of tramp air in the furnace (shown earlier in Fig. 3.11). Any leakage of air into the combustion chamber between the burners and the analyser will seriously disrupt the effectiveness of the measurement as a control parameter for operating the burners. In fact, with a boiler operating with a 1 per cent O_2 level in the flue gases, a 1 per cent leakage of air causes a 20 per cent error in the measured oxygen level.

For these reasons, some boiler control systems combine oxygen trim with a correction based on the opacity of the flue gases—since a reduction of air quantity leads eventually to the production of smoke. However, this concept requires the boiler to be operating badly before the measurement can come into effect. A better approach, made possible through the emergence of sensitive, fast-responding, reliable and accurate CO analysers in recent years, is the use of carbon monoxide analysis (combined with oxygen) to provide effective control of the combustion process.[6]

Carbon monoxide trimming The use of carbon monoxide monitoring to optimize combustion was originally evaluated in the control of gas turbines,[7] but it is equally applicable to the combustion process in boilers. CO trimming has the advantage that it operates on a measurement that is a direct indication of unburned fuel in the flue gases, and therefore it indicates the completeness of combustion. It is also largely independent of the type of fuel being burned and is virtually immune to the effects of any air leaking into the furnace.

Modern CO monitors are sensitive, accurate and reasonably economical, and the technique is therefore widely accepted as being very viable. However, since the objective of CO trimming is to reduce the gas to the point where it can no longer be effectively measured, it cannot be used to control under conditions where excess air must be increased for operational reasons (such as during load changes). For this reason, the more sophisticated systems use a combination of CO and O_2 trimming. Some of these use CO as the controlling parameter while the boiler load is steady, and switch over to O_2 when load changes occur.

Carbon-in-ash monitors Another indication of the efficiency of the combustion process in a plant burning solid fuels is the carbon content of the ash remaining after combustion has finished. Tests at a 600 MW coal-burning unit[8] indicate that a meaningful analysis of this parameter can be achieved by using an instrument employing an automatic dust-sampling system to draw the fly ash from the gas passes in the region of the economizer, and a miniature fluidized bed to burn the sample, the resulting CO_2 content being analysed to indicate the carbon content.

This measurement enables not only the fuel/air ratio to be trimmed but the classifier settings to be optimized as well. Overall, therefore, the technique provides a valuable indication of the efficiency of the combustion process under reasonably stable conditions—when there are no load changes and when the make-up of the fuel has remained constant for some time. However, it should be recognized that the monitor's response to changes in the composition of the fuel involves the plant's pulverizing/burning/gas-transit delays as well as the time constant of the analyser itself (the latter being in the order of 5–17 minutes). Because of these delays, under conditions where the quality and composition of the coal being burned varies from minute to minute, the carbon-in-ash monitor should be relied on only to provide an indication of the average conditions existing over a given time, and any control loop using the measurement should take this factor into account.

3.2.2 A Simple Parallel Combustion Control System

Leaving aside the methods of optimizing combustion, the fundamental requirement of the combustion control process is to adjust the amount of fuel and air to meet the energy demand requirements of the turbine. One technique for achieving this end is to control the two parameters in parallel, as shown in Fig. 3.13. This shows a simple 'parallel' combustion control system for an oil-fired boiler, and in it the master controller output is fed to the oil control valve and to the actuator controlling the air

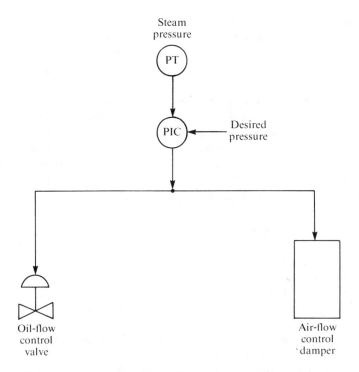

Figure 3.13 Basic 'parallel' control system (for an oil-fired boiler).

flow. Changes in energy demand cause the master controller to alter its output signal, and this results in the fuel and air supplies varying to match the requirement. (Although the diagram illustrates an oil-burning plant, such a system could also be applied in principle to any application where a liquid or gaseous fuel is being used, including gas-fired boilers.)

To obtain the required fuel/air ratio characteristic, it is usual for the fuel valve and the air damper to be 'characterized'—that is, for their movements to be related to each other in a non-linear fashion. This is normally a fixed relationship, since it does not lend itself to being varied by any of the combustion-optimizing systems described earlier.

Parallel control is employed on many smaller boilers, and is quite adequate for such applications. However, in larger installations it becomes necessary to consider the differences between the dynamic responses of the fuel and air systems.

These differences apply whatever the nature of the fuel being burned, but they are particularly acute in the case of boilers burning liquid or gaseous fuels. In such installations, the response of the fuel system to changes in demand is much faster than that of the air system. Consequently a sudden increase in heat demand will cause the fuel supply to increase before the air system can respond, resulting in less complete combustion of the fuel and the production of black smoke every time the load increases. Conversely, sudden decreases in load cause the fuel flow to diminish quickly, yielding momentarily high oxygen contents of the flue gases.

Various ways of overcoming this problem have been employed. In the simplest, a non-linear time-delay function is introduced in the signal to the oil valve, so that sudden increases in demand are held back until the air system has had time to respond, while sudden reductions in demand are immediately passed on to the valve. This concept is simple and can be effective, but it requires the careful adjustment of the time constant to match the dynamic response of the plant. Also, since the plant response does not remain constant over an extended period of operation, the performance of the system will deteriorate as the plant ages, unless the system is regularly retuned to match the current process characteristics.

Another limitation of the simple system is the fact that it relies on open-loop control of the fuel and air supplies. The oil valve is opened in anticipation that the flow will increase accordingly, but if it does not the system cannot react until the shortfall of energy input results in the reduction of steam pressure. (More dangerously, a similar deficiency in the combustion-air supply can admit large amounts of unburnt fuel into the furnace. Accidental ignition of this fuel when it has collected in sufficient quantities will result in an explosion.)

3.2.3 'Cross-Limited' Combustion Control

An ingenious answer to all these problems is represented by the so-called 'cross-limited' system, as shown in Fig. 3.14. Here, separate closed-loop controllers are provided for the oil and air systems. These improve the system's response to losses or failures in these areas, but they also enable the use of minimum and maximum selectors. These selectors, provided at the desired-value inputs of these controllers, operate in the following fashion.

In the steady state, equilibrium is achieved when the master pressure controller output is steady and the oil-flow and air-flow controllers are holding their output signals at values that keep their respective input error signals at zero. In this situation both the oil-flow signal and the characterized air-flow signal (that is, the air flow multiplied by the gain of the ratio control block) are equal to the master controller output.

Should the master controller now demand an increase in the firing rate, its output signal will become larger than the oil-flow and (characterized) air-flow signals. The *maximum-selector* block at the air controller input therefore immediately latches on to the master, while the *minimum selector* at the fuel controller input latches on to the characterized air-flow

MODULATING CONTROL SYSTEMS 61

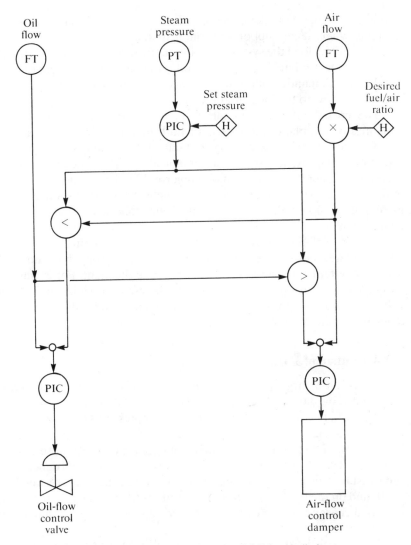

Figure 3.14 Simple 'cross-limited' combustion control (oil-fired boiler).

signal (and ignores the master). The fuel controller therefore initially ignores the increased demand, and begins to act in a 'fuel/air ratio' control mode, adjusting the fuel flow to match the available combustion air supply—which is itself being increased by its own controller to meet the rising demand.

The system therefore calls on the forced-draught fans to increase their throughput at the maximum possible rate, and the fuel flow is brought up to meet the demand only when the air flow has increased to the point where it can support the combustion of the fuel.

On a falling load the system operates in the reverse manner, with the minimum selector latching on to the collapsing master signal and reducing the fuel demand, while the air-flow controller matches the forced-draught air flow to the measured fuel flow.

The cross-limited system alters its response to meet the actual dynamics of the plant, and its use of both fuel and air measurements enables it to cope with a wide range of faults. If, for example, the air flow suddenly collapses, the system automatically reduces the fuel flow—thereby avoiding the production of dangerously rich fuel concentrations in the furnace. Also, if a fault causes the fuel valve to open suddenly, an increased air flow will be demanded in an attempt to maintain safe combustion.

In spite of all this, the failsafe features of the cross-limited system are not complete, and it is necessary to provide supervisory, back-up and interlock systems to ensure the safe operation of the plant. For example, although the control system would react to the sudden opening of the fuel valve, the slow response of the draught plant would still result in a dangerous build-up of fuel-rich gases in the furnace.

3.2.4 Control of Larger Systems

In practice, on large plants the simple combustion control systems discussed above are rarely encountered in the form described. In particular, the burning of coal involves the use of complex subsystems.

Multiple burners require careful consideration, since simply controlling the flow of fuel to all burners will not guarantee that safe and efficient conditions exist at each one. Flows to an individual burner will depend on the resistance offered by the fuel supply lines and the burner assembly itself, and on the vertical position of the burner (due to the differences in potential energy between burners at the bottom of the furnace and those at the top).

Because of these factors, simple control of the overall fuel and air flows will result in more fuel being admitted to some burners and less to others, with the result that some burners will be fuel-rich while others will be fuel-starved. Unfortunately, any surplus fuel at one burner will not necessarily be burned when it enters the furnace: in the complex flow patterns of a large combustion chamber there is always a risk that unburned fuel will collect in one area or another, awaiting sudden—explosive—ignition when the right conditions arise.

It might appear that the only way of ensuring safe and efficient combustion is to measure the fuel and air flow *at each burner*, but this ideal is rarely practicable. Fuel-flow meters are not inexpensive and air-flow measuring devices often require the provision of long lengths of straight duct. Bearing in mind the volumes of air being handled, the duct cross-sectional areas are necessarily large, and since the measurements will demand reasonably laminar flow—a condition that demands that the air has travelled along a straight portion of duct whose length is directly related to the cross-duct dimensions—the dimensions of the overall primary element complex become correspondingly enormous. (Nowadays it is becoming increasingly common to use hot-wire anemometers for air-flow measurement, and these do not require the flow to be quite so laminar. However, the accuracy of the measurement still depends on the degree of turbulence, and so the argument given in the text still applies, albeit to a lesser extent.)

Various attempts have been made to solve these problems, some employing actual fuel and air measuring devices, while others use inferential techniques where the burner is itself used as a calibrated orifice, and the air and fuel flows are inferred from the differential pressures across it.

3.2.5 Combustion Control in Fluidized-Bed Boilers

The combustion process in fluidized-bed boilers is significantly different from that of conventional plant, and theoretically the control system should recognize this fact. However, in practice, the most significant differences between the control systems of fluidized-bed and conventional boilers will be found in the air supply arrangements, since in most cases the fuel feed is related to the steam flow.

3.3 PULVERIZER CONTROL

The objective of controlling a coal pulverizer (or 'mill') is to regulate the amount of coal being ground in such a way that enough of the fuel is fed to the burners to meet the load requirements, while maintaining optimum operation of the pulverizer itself (that is, not starving it, overfilling it or allowing it to overheat).

The coal must also be dried, so that the air supply must be warm—the wetter the coal the higher the necessary air temperature.

In general, therefore, the control systems for pulverizers have three components:

1. A system that adjusts the throughput to meet the boiler's fuel demand.
2. A system that regulates the ratio of coal to air input to maintain the correct inventory of coal in the pulverizer.

3. A system that adjusts the temperature of the air entering the pulverizer, so that the coal is adequately dried without overheating the fuel/air mixture leaving the pulverizer (which would run the risk of the fuel igniting before it reached the burners).

Four main categories of pulverizer are considered here:

- Pressurized vertical-spindle types
- Suction-type vertical-spindle pulverizers
- Pressure-type horizontal tube pulverizers
- Suction-type horizontal tube pulverizers

In modern plant the suction-type pulverizers are seldom encountered, but they are described here since many older installations still use them.

3.3.1 The Control of Pressurized Vertical-Spindle Pulverizers

The basic system employed in controlling a pressure-type vertical-spindle pulverizer is shown in Fig. 3.15. Here, the demand for fuel is used as a desired-value signal for the primary air controller, which compares the demand with a signal derived from the differential pressure across a restriction in the primary air duct and adjusts the fan delivery accordingly. The primary air transports the coal into and through the pulverizer, and so the amount of fuel delivered to the burners is related to the primary air flow. (In practice, the square-law relationship between the differential pressure measurement and the primary air flow is taken into account by a square-root extractor at the primary air controller measured-value input. Alternatively, a linearized flow measurement can be used in place of the differential pressure signal.)

This part of the system operates on the assumption that the coal fed into the pulverizer will match the demand—a function that is achieved by a second controller (item 2 on Fig. 3.15), which regulates the coal feeder speed to maintain a predetermined relationship between the primary air differential pressure signal and the pressure drop across the pulverizer. If too much coal is delivered by the feeder, the pulverizer will tend to fill up with coal. The resulting increase in the pressure drop across the pulverizer will be corrected by the pulverizer differential controller.

Clearly, the pressure drop across the pulverizer will depend not only on the amount of coal in the pulverizer but also on the amount of air flowing through it. Over the operational range of the pulverizer, there is an optimum relationship between the primary air differential pressure and the pulverizer differential. This relationship, known as the pulverizer 'load line', is defined by the manufacturers but it is determined by the exact geometry of each pulverizer and by variations due to wear. For

MODULATING CONTROL SYSTEMS

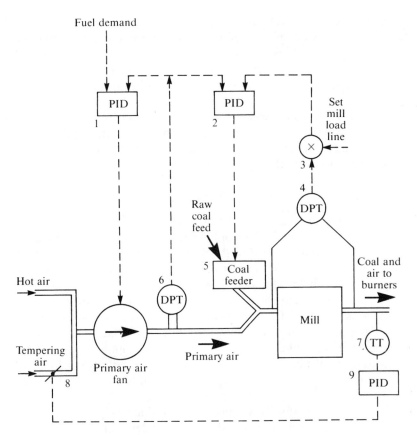

Figure 3.15 Control of pressure-type vertical-spindle mill.

these reasons it is usually adjusted on site—and readjusted from time to time—as operational experience is obtained of the actual characteristics of the pulverizer and the associated plant at the time. In certain situations, pronounced non-linearities are defined or discovered in the load line, and these require the use of non-linear function generators to obtain the appropriate characteristic.

The dynamic response of this type of pulverizer is determined very largely by the classifier characteristics and the throat gap. Since these parameters are affected by wear, the response of any pulverizer will be different from the others on the same plant, and will vary with time.

In order to dry out any residual moisture in the coal, the air supply for the primary air fan is derived from the air duct after the boiler's air heaters. However, it is necessary to ensure that the temperature of the coal/air mixture is kept to a safe value, to prevent premature ignition. This function is achieved by the final subsystem of the pulverizer control complex, a temperature controller (item 9 on Fig. 3.15), which mixes cool 'tempering' air with the heated air to obtain the correct temperature. Although Fig. 3.15 shows this controller acting on a single damper in the tempering air duct, it is common for two sets of dampers to be modulated in opposition—one in the hot air duct, the other in the tempering air stream.

In order to ensure safe operation, the design of this loop has to balance the need for fast response against the conflicting requirement for a measurement device that is adequately protected against the severe erosion that occurs in the location of the temperature measurement.

3.3.2 The Control of Suction-Type Vertical-Spindle Pulverizers

With suction pulverizers, a fan draws the coal/air mixture through the pulverizer, and the main control loop (illustrated schematically in Fig. 3.16) acts to vary the speed of this 'exhauster' to meet the demand for fuel. (Alternatively, the flow is sometimes controlled by means of a damper at the exhauster inlet.) Again, a closed loop compares the demanded fuel with the measured air flow into the pulverizer.

One advantage of a plant using variable-speed exhausters (rather than dampers or vanes) to control the flow is the fact that the system resistance is effectively constant at all throughputs. This allows a measurement of motor power to be used as an indication of the total throughput (coal and air), though obviously this technique cannot be used where variable-speed drives are fitted.

Pulverizers of this type operate most effectively with a defined suction head, and this is achieved by a controller (item 7 on Fig. 3.16), which adjusts a damper in the air duct to maintain the suction at the optimum value. In some installations the setpoint for this loop is derived from the boiler load via a non-linear function generator.

As with the pressure-type pulverizer, a separate loop controls the admission of tempering air to hold the temperature of the coal/air mixture at a safe value.

3.3.3 The Control of Horizontal-Tube Pulverizers

The general principles for controlling horizontal-tube pulverizers are very similar to those of the corresponding type of vertical-spindle pulverizer (that is, pressure or suction type). However, the method of measuring the coal content of the pulverizer is different: in a horizontal tube pulverizer it is derived from the pressure differential between two tubes, one of which is inserted into the bed of coal at the bottom of the pulverizer, the other at the top. (Another technique is to use an acoustic measurement, since the coal quantity present in the pulverizer is related to the noise generated by the pulverizer action.) However it is derived, the coal-level signal is used to vary the speed of the coal feeder.

MODULATING CONTROL SYSTEMS **67**

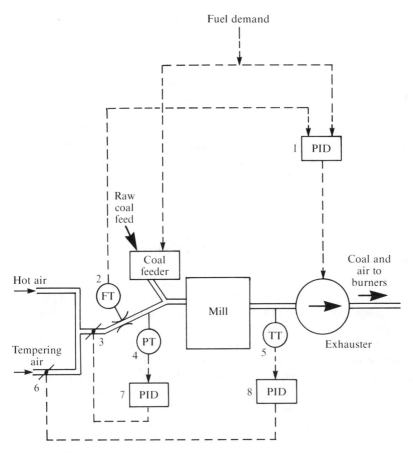

Figure 3.16 Control of suction-type vertical-spindle mill.

Like the pressurized vertical-spindle version, this type of pulverizer is provided with a sealing air supply, which is controlled to maintain a fixed pressure within it.

3.4 MIXED-FUEL FIRING

Many boilers are designed to burn a mixture of fuels. In general, the control principles involved in these systems require that the fuel flows are first of all converted to a common baseline (for example, converting the oil flow to its coal equivalent in a plant burning oil and coal) and then added together to produce a total fuel-flow signal, measured in terms of one of the component fuels.

Operational requirements determine the form of control system that will be used. For example, it may be necessary to burn one fuel

68 BOILER CONTROL SYSTEMS

preferentially, or to limit the amount of firing of a fuel. Each case must be analysed on its own merits, and the control strategy dictated accordingly. However, experience with such mixed-fuel systems has shown that extreme care must be taken in designing the systems to prevent the possibility of overfiring or operating with dangerous fuel/air ratios. In particular, the boiler safety valves are designed to handle 100 per cent of the evaporation in the event of the turbine inlet valves shutting, and if the total fuel capacity is equivalent to more than 100 per cent, the safety valves will be unable to relieve the pressure in the event of a trip while the combined fuels are being burned. In such an event the boiler pressure will continue to rise until the plant explodes. The only way to avoid this would be to fit additional or larger safety valves—which would be much more expensive than including the correct interlocks and controls.

3.4.1 Problems Associated With the Control of Multiple Pulverizers

Boilers provided with more than one pulverizer lead to several problems of control, one of which is associated with the task of transferring from manual to automatic control.

Automatic control systems on hazardous plants such as boilers require the provision of manual override facilities, so that the human operator can intervene if a fault in the system risks the safety of the plant. Usually, this manual override takes the form of a 'hand/auto' (or 'auto/manual') station, which enables the controller to be switched off and its commands replaced with manually generated signals. A simple version of such a system is shown in Fig. 3.17a, where the command signal (S_o) for the actuator can be switched from the controller signal (S_a) to a manually generated signal (S_h). With analogue or digital controllers that generate continuous output signals (i.e. not pulsed outputs), means must be provided to make the automatic and manual signals equal at the instant of transfer from hand control to automatic—otherwise dangerous transient disturbances will be introduced into the plant at the instant of switch-over.

One technique—shown in Fig. 3.17b—links the actuator demand (S_o) to a so-called 'reset' connection at the controller. The signal applied to the reset terminal (now called S_r) forces the controller output to follow the final command signal while the loop is on hand control, so that the switch-over from manual to automatic control is smooth.

Because this type of system requires no manual intervention to balance the signals before switching from manual to automatic control, it is commonly referred to as a 'procedureless, bumpless transfer' system.

This configuration is well established practice throughout the process control industry, but it works only with single loops: difficulties arise

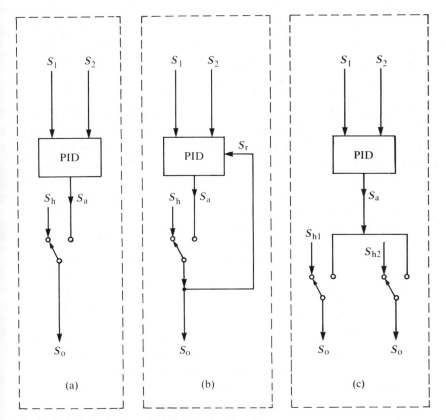

Figure 3.17 'Bumpless' transfer—problems with multiple outputs.

where one controller regulates two or more actuators in parallel—a feature of boilers with multiple pulverizers.

A simple example is shown in Fig. 3.17c, where one controller positions two actuators. When the loops are on automatic control, all outputs will be nominally equal, but when they are under manual control, the operator may elect to run the plant with the two signals at significantly different levels (say, with S_{h1} at 100 per cent and S_{h2} at 10 per cent). There is no way of predetermining which of the hand/auto stations will be the first to be switched to automatic, and it is therefore not possible to tell which of the outputs should be connected to the 'track' terminal in order to provide rapid bumpless transfer between control modes.

System vendors who have not encountered multiple-output control configurations often fail to recognize the difficulty of pulverizer 'first-to-auto' transfer, and some have been known to offer 'quick and easy' solutions that appear to work in theory but invariably fail to do so in practice. More experienced manufacturers know the reality of the problem, and have adopted sophisticated and field-proven techniques, such as applying a transient delay when transferring the first pulverizer to auto, allowing the control output to be 'forced' to take up the value of the relevant manual status. Whatever solution is offered, however, it is important that the means of obtaining multiple-output control is carefully considered—and preferably seen in operation on a power plant—before accepting a proposed configuration.

Another problem of controlling a boiler with multiple pulverizers (or burners, or feed pumps, for that matter) is the need to keep the loop gain constant as the number of operational control outputs is altered. Clearly, a given change in the demand from the pulverizer master will produce more heat change if eight pulverizers are in service than it would if only one is operating. It is therefore necessary to apply gain adaptation techniques in order to stabilize the loop as the operator changes the number of pulverizers running under automatic control.

Yet another problem inherent in the use of multiple pulverizers is the complexity of the information that is presented to the operator. Because of the multiplicity of complex loops around each pulverizer, the control console would become crowded and difficult to manipulate if all the control parameters, all the hand/auto stations and all the manual control facilities for all pulverizers were to be made available to the operator.

A practical solution is to *multiplex* the pulverizer controls and indications. This means that facilities for only, say, two pulverizers are made available at any time, although provision is made for the operator to select *which* two are to be used at any time.

This approach is ergonomically efficient, since it is more 'user friendly' in that it offers the operator only the facilities that it is practicable to handle at any one time and does not deluge him or her with unnecessary information or facilities. However, multiplex operation leads to some extremely complex arrangements of software, cables and interlocks, and must be very carefully engineered if it is to be safe and effective.

3.4.2 A Typical Pulverizer Control System

Figure 3.18 shows part of the control systems of a vertical-spindle pressurized pulverizer (one of eight on a 660 MW boiler burning pulverized coal with the ability to meet some of the load demand by firing heavy fuel oil).

The firing demand signal is fed in parallel to each of the pulverizer groups via a master hand/auto station (which enables the operator to control all the operating pulverizers from a single station). This signal is transmitted via a gain-control block, which compensates the loop gain for alterations in the number of pulverizers in service.

Each pulverizer feeds fuel to a group of burners, and to give the operator facilities with which to control any one such group indepen-

MODULATING CONTROL SYSTEMS 71

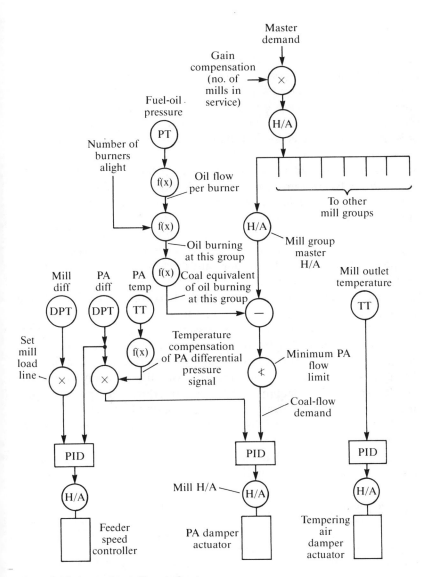

Figure 3.18 A complete mill control system.

dently of the rest, an individual hand/auto station is provided for each of the eight groups.

The output of this hand/auto station represents the total fuel demanded from the group but, as in this case the boiler is designed to burn oil as well as coal, a signal representing the coal equivalent of the oil flowing to the burner group is subtracted from the fuel demand. The resultant signal:

(Total fuel demanded from group) − (Fuel oil flowing to group)

represents the throughput demand from the pulverizer, and it therefore forms the desired value for a controller that adjusts a damper to keep the primary air (PA) differential pressure at the correct value. A minimum-limit block prevents the system shutting down the pulverizer throughput to a value where flame instability could arise, while the PA differential pressure is compensated against the temperature variations that occur at this point.

The feeder speed is modulated to keep the differential pressure across the pulverizer in step with the PA differential, according to the pulverizer load line. The final loop maintains the temperature of the coal plus air mixture leaving the pulverizer at a safe value.

The fact that this system is designed to operate with mixed-fuel firing (coal and oil) leads to special considerations. Depending on the design of the plant, it may be possible to burn any mixture of fuels (up to 100 per cent of each) under automatic control, but in this case it would be necessary to provide some form of interlock to prevent *both* fuels being burned simultaneously in such a way that the boiler was overfired. Practically, although it would be possible to burn both fuels under automatic control, such a *modus operandi* would lead to quite a complex design of system (with, among other things, compensation for the different loop gains when burning coal or oil). In many cases it is simpler—and quite adequate—to provide interlocks that prevent operation with both fuels under automatic control simultaneously.

This group of systems illustrates many of the complexities of boiler control, some of which are summarized in Table 3.1.

It should be noted that the systems shown in Fig. 3.18 are highly schematic, and require the addition of several enhancements in order to

Table 3.1 Some complexities of boiler control.

Feature	Implication
Multiple pulverizers	'First-to-auto' control tracking Gain compensation for the number of pulverizers operating under automatic control
Multiple fuels	Interlocks to prevent overfiring Precautions to allow for different characteristics of fuels
Multiple sub-loops	Multiplex operation (to avoid cluttering the operator's console with too many controls and too much information)

MODULATING CONTROL SYSTEMS 73

yield a practical set-up. Also, they comprise only the modulating section of the control philosophy. Complex logic is also needed to provide sequenced start-up and shut-down of the pulverizers, and to ensure the safe operation of the plant.

3.5 CONTROL OF AIR FLOW

Because most of the power-plant boilers we are considering would be provided with two forced-draught fans, and because these deliver a compressible fluid (air) into a single enclosure (the furnace or the furnace/windbox combination), various control problems arise due to the interaction between the systems. For example, when the delivery from one fan is altered, an alteration is produced in the pressure of the discharge duct. This duct is common with, or linked to, the discharge duct of the other fan, whose throughput therefore also tends to change.

3.5.1 Parallel Operation of FD Fans

On the simplest level, a basic loop can be constructed with a single controller modulating the two fans in parallel (as shown in Fig. 3.19). At

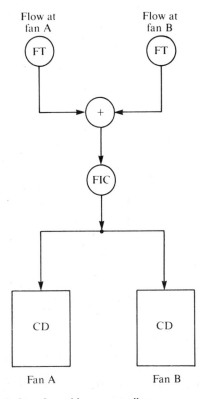

Figure 3.19 Control of two fans with one controller.

first sight, however, it might seem that this would present the same fundamental problem that is encountered when multiple mills are regulated by a single controller—that is, it requires the loop gain to be adapted according to whether one or both of the fans is under automatic control, as well as the provision of 'first-to-auto' hand/auto transfer facilities.

In fact, the latter consideration is irrelevant, since the system is inherently simpler, with no subsystems to prevent the use of simple techniques for ensuring smooth transfer between manual and automatic control. It also transpires that the problem of gain compensation is comparatively simple, since there are only two relevant states of the system:

- one fan on auto (the other being off, or on hand control);
- both fans on auto.

Therefore only two switched gain settings are needed.

Another apparent deficiency of this simple system would appear to be the fact that it incorporates no facility for ensuring that the through-

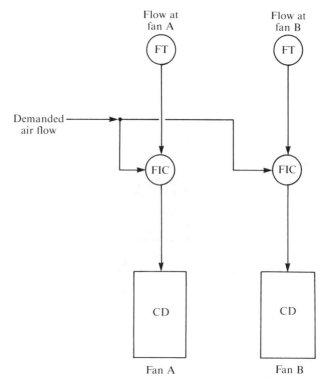

Figure 3.20 Control of two fans with individual controllers.

puts of the two fans are equal. On this basis the arrangement shown in Fig. 3.20 would appear to be better for this purpose: it uses individual controllers for each fan, sharing a common desired-value signal so that, as the air-flow demand varies, each fan is commanded equally to respond to the change.

In practice, although such systems have been used, they are extremely difficult to tune and optimize, and are easily disrupted by deterioration of the plant items with the passage of time. The difficulty of tuning is due to the interactions mentioned at the beginning of this section, which result in one fan's controller acting to counteract changes in flow resulting from actions taken by the second fan's controller—whose own controller then responds in an attempt to eliminate the change.

This interaction leads to 'hunting'—cyclic instability in the operation of both fans—which is operationally disturbing, inefficient and possibly unsafe. The single-controller approach is therefore preferable.

3.5.2 The Air-Flow Demand Signal

For the most efficient operation of the plant, the signal that dictates the quantity of air delivered at any instant must be related to the amount of air theoretically needed to burn the fuel flowing to the burners at that time. In a boiler burning a liquid or gaseous fuel, this relationship can be derived from a fuel-flow measurement, but in a boiler burning a solid fuel (or a combination of solid and fluid fuels) the use of a fuel-flow measurement is not practicable.

The measurement of coal flow suffers from several problems, two of which are as follows:

1. Although the combustion control system must receive a signal representing the weight of the *combustibles* in the fuel, practical weighing devices measure the *total* weight of the fuel (including, along with the combustibles, impurities such as the moisture and ash content of the fuel). Currently, the separation of the useful element from the total is not possible.
2. Weighing systems capable of handling the large throughput of pulverized fuel are at present not sufficiently accurate or fast-responding for control purposes.

Furthermore, measuring the flow of coal anywhere other than at the burners themselves necessitates taking into account the transport time of the flow from the point of measurement to the burners and in some situations the mill inventory of coal.

These problems make it difficult, if not impossible, to obtain a sufficiently accurate, fast-responding and meaningful measurement of

76 BOILER CONTROL SYSTEMS

the flow of coal, and for this reason most systems for boilers burning this type of fuel use an inferential signal—and the parameter used for this purpose is steam flow.

Controlling the air flow to maintain a defined steam-flow/air-flow ratio is well established and, in the absence of a viable means of measuring coal flow, fairly satisfactory.

Again, owing to the multiplicity of burners in large boilers, difficulties arise in distributing the air flow to individual burners, and further problems arise when a mixture of fuels is being burned. (The use of the more sophisticated gas analysis systems to alleviate these problems was described in Sec. 3.2.1.)

A popular solution, shown in Fig. 3.21, is to regulate the air pressure in the windbox with respect to the steam flow, since the optimum pressure

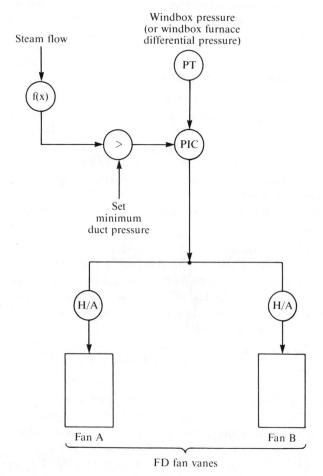

Figure 3.21 Windbox pressure control system.

at that location for a given coal flow will be related to the boiler load. As soon as another burner is introduced, the additional air flow through its registers will cause a momentary drop in the windbox pressure, which the system will counteract by demanding more air from the FD fans.

In the system shown, an additional sub-loop prevents the windbox pressure from falling below an operator-set value: by setting this value to a high level, the operator can effectively alter the system configuration so that it too maintains a constant hot-air pressure at all loads. (This is typical of a system design approach that recognizes the fact that the parameters defined by the boiler designer may need to be adapted in the light of practical operating experience with the actual plant.)

The system takes account of the air flow used to cool idle burner registers since this contributes to the total amount of air flowing into the combustion chamber, although as already explained this does not mean that combustion conditions at each burner are optimized.

The objective of maintaining a safe level of excess air is, of course, relevant to boilers burning conventional fuels. This requirement must always be balanced against the need to minimize the amount of excess air, but where unconventional fuels or fuel combinations with high sulphur contents (such as coal/water mixtures or bitumen/water mixtures like Orimulsion[9]) are being burned, additional factors may make it particularly important to reduce the excess air to the minimum possible value in order to avoid excessive corrosion and unacceptable flue-gas emissions. A factor to be considered is whether there are any other elements present in the fuel and whether these require special consideration because of the chemistry of the combustion process. For example, vanadium has oxides that act as catalysts for sulphur, leading to increased sulphuric acid formation in the combustion chain.

3.5.3 Air-Flow Control in Fluidized-Bed Boilers

The fact that the combustion control system of any plant must be related to the characteristics of the fuel being handled is as true for a fluidized-bed boiler as it is for a conventional one. Therefore, the admission of fuel in a coal-burning plant of this type is regulated in parallel with the supply of air to meet the demands of the boiler master, while in an oil-fired version it is based on the familiar cross-limited principle.

However, the control of the combustion process in these boilers has to take into account the air that is necessary for the fluidization process itself, the quantity being dominant at lower loads and becoming progressively less significant at higher loads. The excess air will therefore vary significantly—but quite properly—over the operational range of the plant, and the control system should recognize this characteristic and apply a 'desired O_2' level which is varied over the load range.

Also, as it is usual to supply air at different points in the combustor, each of these quantities must be measured—although if a fixed supply of air is used in any stage of the process, a fixed signal may be used to represent this parameter.

The air-flow control system must also recognize the special characteristics of the bed design. For example in one type of multi-solids fluidized-bed (MSFB) boiler,[10] in order to keep good NO_x control, the air supply to the lower section of the bed is kept well below the theoretical stoichiometric fuel/air ratio, with additional oxidizing air being supplied to the upper region of the bed via a secondary air fan. The throughput of this fan is controlled to meet the load-related flue-gas oxygen content criterion mentioned earlier.

In all the recognized commercial designs of fluidized-bed boiler, the bed level is inferred from a measurement of the differential pressure across the relevant combustor zone, and controlled by regulating the amount of air being fed to that zone.

The temperature of the fluid bed is measured by means of infrared pyrometers, special precautions being taken to obtain an average reading of the temperature across the whole bed. In the Battelle design of MSFB plant, the bed temperature is controlled by modulating non-mechanical air valves, which recycle solids through the various zones of the combustor.

Since a major advantage of the fluid-bed concept is the reduction in undesirable emissions produced by the boiler, it is usual to measure the SO_2 content of the flue gases and to modulate the quantity of limestone being fed into the combustor to keep these at the optimum value. This reduces the SO_2 emission while minimizing the consumption of limestone.

3.6 FURNACE PRESSURE CONTROL

In many respects, the control configuration for the induced-draught fans is similar to that of the forced-draught fans: a single controller modulating the delivery of the two fans in parallel. The reasons for using a single controller are the same as those applying to the FD fan system: the fact that the two fans are connected with a single entity (the furnace).

In furnace pressure control systems, the objective is to maintain a fixed pressure (always a slight suction) within the combustion chamber. Because of the large size of the furnace, it is common to measure the pressure at two points and to work on the average measurement so derived. It is important to take into account the variation of suction up the height of the furnace (a modest suction at a high point will correspond with a greater suction lower down). Therefore, the two tapping points must always be at the same horizontal level.

Since the suction/height relationship will vary with load and possibly even with time, it is not possible to allow for a fixed gradient, and the tapping points must therefore be located at a high level in order to avoid the emission of dust from the furnace caused by operating at a positive pressure lower down.

The large physical dimensions of boilers imply that even a very slight suction imposes large forces on the side walls, and the boiler's ability to resist these forces is clearly defined by its design. In many installations, therefore, it is common to see a subsystem that acts to protect the plant against excessive pressures.

Because of the interaction between all the parameters in a boiler system, various attempts have been made to apply feed-forward techniques to the control loops—and the furnace pressure system has seen more than most of these.

There are good reasons for this. The induced-draught fans of a boiler are large, and the quantity of flue gases they extract is enormous—factors that do not lead to rapid response. It would therefore be very useful if some means could be found to programme the ID fans so that they start to respond as quickly as possible, and there is no lack of availability of apparently suitable forewarning signals.

In the simplest examples, the master pressure signal is used to give advance indication of impending gas throughput changes, in an attempt to prepare the system against the resulting furnace pressure fluctuations. Alternatively, the air flow at the FD fan delivery has been used for the same purpose.

In practice, these systems have proved to be very difficult to commission successfully, and those that have been set up to work in a reasonably effective manner are reported to be prone to drift with time, so that plant maintenance personnel usually detune (i.e. switch off) the feed-forward signal.

In attempts to overcome these failings, control engineers have adopted 'd.c. blocking' techniques, which feed forward only those changes which are likely to be significant. However, it is common to find that, even then, many such systems have been set up by commissioning engineers so that the feed forward is effectively removed (probably because of a lack of time—during the intense pressures of commissioning the plant—to adjust accurately the d.c. blocking parameters).

Figure 3.22 shows the furnace pressure control system of a coal-fired 660 MW boiler in a schematic form. It includes many of the features described above, including duplicated pressure transmitters to obtain the average pressure measurement, parallel operation of the two ID fan

80 BOILER CONTROL SYSTEMS

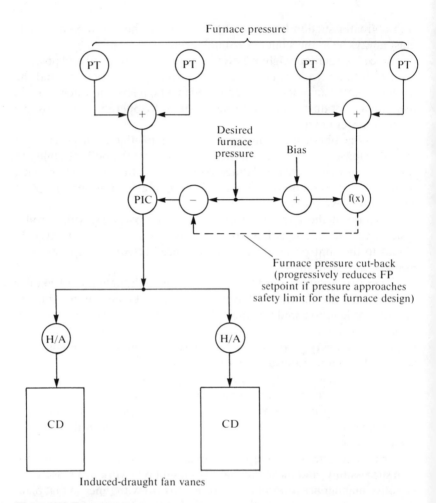

Figure 3.22 Furnace pressure control system.

vanes and a pressure cut-back system. The last of these subsystems uses, in this example, transmitters (also duplicated) that are completely separate from those used for normal control. This sounds an expensive luxury, but it ensures that the plant cannot be endangered by failure of a transmitter. The cut-back system compares the averaged signals from these transmitters with a signal that is set slightly above the setpoint of the normal pressure controller. If this signal is exceeded by the measured pressure, the cut-back system begins to reduce progressively the actual setpoint for the modulating controller.

In practice, the initiation of any overriding action like this is normally arranged to trigger an alarm, which will draw the operator's attention to the fact that a dangerous condition has been detected and is being corrected by the system. This encourages the operator to analyse the situation to see what could have caused the problem, and to take any other remedial action that may be necessary.

3.6.1 Furnace Pressure Control in Boilers With FGD

In many cases, flue-gas desulphurization (FGD) plant has been added to existing boilers in order to comply with emission-control requirements that did not exist when the plant was originally conceived. Additional 'booster' fans are therefore provided within the new plant to overcome the additional draught losses imposed by the FGD process, and these fans are controlled to compensate for the resistance of the FGD plant over the load range.

This objective is generally accomplished by controlling the booster fans to maintain a predetermined desired pressure at the ID fan outlets. The desired pressure is set to a level that corresponds to the situation which existed before the addition of the FGD plant.

The objective of the control philosophy is to decouple the FGD plant effectively from the boiler. However, the booster fans and the ID fans operate in tandem, and to complicate matters further the dynamic performance is affected by the action and response of another set of fans, the FD ones.

It is therefore necessary to take steps to ensure that the resultant system will be stable under all operational conditions, and that no conceivable circumstance can cause undesirably high furnace pressure transients or disturb the combustion process to the point where unstable conditions could arise.

In summary, the design of the FGD control system should bear in mind the interactions between the FD, ID and booster fans, and it should be such that the overall plant achieves the optimum levels of performance and safety.

3.7 FEED-WATER CONTROL

The primary objective of a boiler feed-water control system is to match the supply of water to the evaporation rate. With once-through boilers, the final steam temperature is dictated by the total flow rate of water into the system (that is, feed water *and* attemperator spray water) and the heat energy input to the furnace, and the feed-water control strategy has to take this into account. In drum boilers, on the other hand, the most important requirement is to maintain the level of water in the drum at around the mid-point: an excess of feed causes the level to rise, while a deficit causes it to fall.

Whatever the type of plant, the feed-water flow is regulated either by means of modulating the control valves or by varying the speed (and hence the delivery rate) of the pumps—or by a combination of both.

In drum boilers with spray attemperators, water from the feed pumps is also used to control the steam temperature, and it is necessary to maintain an adequate margin of spray-water pressure above the prevailing steam pressure. The pressure drop through the boiler's water and steam passages is itself inadequate for this purpose, and providing the necessary differential pressure conditions is one of the functions of the feed regulating valves. A common control configuration uses the feed valves to control the drum level, while the feed-pump speed is varied to maintain the desired differential pressure across the valves.

This philosophy contributes to optimizing the efficiency of the plant by maintaining the pumping energy at the lowest possible level consistent with keeping adequate pressure available for spraying at the attemperators. It also provides rapid response to drum level changes and steam/feed mismatches and maintains ideal operational conditions for the feed regulator valves.

It is possible to set the desired value of the feed-pump speed controller so that a fairly large pressure drop is held across the attemperator spray valves. This saves the feed pumps from having to deal with short-duration steam temperature changes, since these will be largely dealt with by the valves. However, this *modus operandi* results in some reduction of efficiency and greater wear of the valve internals as a result of the large pressure drop across them.

Such control systems (regulating the drum level via the feed valves and using the pumps to maintain a controlled differential pressure across the valve manifold) are not very easy to commission, because the feed valve and feed pump loops are mutually interactive. A lower level of interaction is achieved by a system (used in some plants) where the drum level controller varies the pump speed and the valves are modulated to keep a constant differential pressure across themselves. This option also offers more consistent regulation of the spray-water differential pressure—and therefore assists in achieving better steam temperature control—since the feed-valve closed loop is a very fast-acting one.

With once-through boilers the control strategy is more complex, with the pumps usually assuming the major control rôle, leaving the valves to be used primarily for pump protection. The systems for achieving this are discussed later.

Irrespective of the plant type, in the larger installations the feed and steam paths through the boiler are not usually single streams. It is usual for several feed valves to be provided in a single or duplicated manifold arrangement, and to have several parallel steam paths through the various evaporating, superheating and reheating passages and the turbine stages. Apart from making fabrication more easily feasible, these measures provide a level of redundancy. For example, a unit with three feed valves—each rated to carry 50 per cent of the maximum throughput of the boiler—can be operated at full load with any one of the valves out of action.

Furthermore, it is common to have a combination of electrically- and steam-driven feed pumps. In such cases, the configuration may comprise, say, two 50 per cent duty electric pumps and one 100 per cent duty steam-driven pump.

For reasons of simplicity, the following system descriptions are related to single feed and steam paths: in practical installations, means have to be provided for balancing the flows through parallel paths, transferring from one pump to another, automatically bringing a pump or valve into service if a running one fails, adapting the loop gain for changes as more auxiliaries are brought into service, and so on.

3.7.1 Level Control in Drum Boilers

In drum-type boilers, although the level of water in the drum must be controlled by regulating the feed flow more or less in step with the steam flow, it is not only the steam/feed balance that affects the level of water. Among the other factors that affect this parameter is the effect of load change on the steam pressure. An increase in demand will cause the pressure to fall, reducing the saturation temperature and increasing the boiling rate, with the result that the evolution of steam bubbles in the water increases, causing the level to rise. This process is known as 'swell', and its opposite counterpart—occurring during load drops—is 'shrinkage'.

Both swell and shrinkage have the effect of confusing simple control systems that rely on measuring only the drum level (the so-called 'single-element' loops). Remembering that a load increase reduces the saturation temperature, which in turn causes the level to *rise*, it will be apparent that a level controller will respond by *reducing* the flow of feed water—at the very time when, in fact, more flow is needed.

This phenomenon has given rise to the development of multi-element feed-water control systems. 'Two-element' loops combine the use of a steam-flow signal with the drum level measurement to help the system respond correctly to load changes—for example, the rising steam flow of a load increase compensating for the swell effect and opening the feed valves.

Two-element control is capable of handling reasonable load changes, but a better method of control is to combine drum level and steam flow

84 BOILER CONTROL SYSTEMS

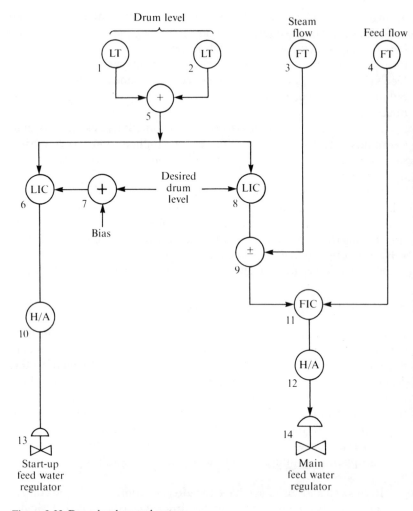

Figure 3.23 Drum level control system.

with a third measurement, feed-water flow. Such 'three-element' systems have long been established for the control of drum boilers handling severe load changes.

Figure 3.23 shows a typical *valve-maintained* drum level control system as applied to modern drum-type boilers, and some of the system's features are discussed below.

Level measurement The discussion of swell and shrinkage will have given a warning that the measurement of drum level, in theory a simple matter, is in reality a complex subject in its own right.

For a start, since level measurements in high-pressure vessels are generally based on the pressure differential existing between the steam in the upper part of the drum and the water below it (the latter being the sum of the pressure and the water head at the tapping point), they are affected by changes in the density of the water. The drum level transmitter must therefore be compensated for variations in the steam pressure within the drum and for changes in the temperature of the feed water.

Then, if the measurement is taken from tappings outside the drum, it should be compensated for the difference between the density of the hot water inside the drum and the cooler water in the reference leg outside.

Finally, various events (such as burner light-off) can cause localized surges through the 'riser' tubes, which are influenced more directly by the event—which will result in the level in that area rising above the mean level. With the large sizes of modern boilers, the storage capacity of drums is small in relation to the throughput of the plant—it is quite normal for a drum to hold only enough water for 10–15 seconds of safe operation of the plant following loss of feed. The accuracy of the level measurement (not to say the reliability of the system) is therefore quite critical. To ensure that an accurate representation of the mean level along the length of the vessel is used, most systems employ two level transmitters, situated at opposing ends of the drum.

One interesting and novel aspect of boiler drum level control is the use of conductivity probes to provide the basic level measurement.

The Hydrastep instrument[11] (originally developed by the UK Central Electricity Generating Board and manufactured by Schlumberger Industries, Transducer Division) gives a very accurate indication of the drum level without being affected by pressure and temperature changes. It consists of a side-arm pressure-vessel water-level-indicating column into which are fitted, typically, 14 or 16 electrodes. The pressure vessel is attached to the boiler such that half the electrodes are above, and half below the normal water level. The instrument relies on the resistance measured between the electrode tip and the indicating column wall, the resistivities of water and steam differing substantially from each other (by over 200:1).

As the water level inside the indicating column rises or falls in unison with the level in the drum itself, so the changes in resistivity are sensed by the electronics, digitally processed and displayed locally and in the control room by means of a red and green columnar LED display. The system employed in conjunction with this instrument compares each electrode channel with the adjacent ones. Since water cannot exist at a level above steam, any such indication must imply a fault. Such errors are flagged and the system allows boiler operation to continue. The system also copes with power supply failures by using duplicated supplies. One power unit supplies the even-numbered electrodes, the other the odd-numbered ones. If one supply fails, the second will permit six equally spaced electrodes to remain in operation, fulfilling the (UK) legal obligation of indicating a 20 per cent water-level change to the operator.

Control during unit start-up and shut-down Under low-load conditions, the operation of three-element systems can be confused by errors in the steam and feed flow measurements (these errors becoming more significant at low flows). For this reason, while control at high loads is via a full three-element system, under start-up conditions a smaller valve (item 13 in Fig. 3.23) is modulated by a single-element controller (item 6), which responds only to variations in the drum level.

This controller shares its setpoint with the three-element controller, but biased to a slightly lower value. In this way, transfer from single-element to three-element control is made effectively automatic. When the unit load has risen to the point where the main feed-water regulator can be brought into operation, the single-element controller is effectively overridden, since the main controller—judging the level held by the single-element system to be too low—will open the main valve to bring the level up to its own desired value. At this point the single-element controller will close down the start-up valve in an attempt to bring the apparently high water level back down to a safe value. The start-up valve will therefore remain shut until the main regulators begin to close (for example, at unit shut-down), when the controller—detecting the falling level—will open up the smaller valve.

Although this system opens and closes the start-up valve semi-automatically, with no other provision to prevent such a condition arising, it can result in the valve being held closed for extended periods with a high differential pressure across it. Since the start-up valve is not normally of the tight shut-off type, such operation will quickly erode the seat, and for this reason the isolating valves upstream and downstream of the start-up regulator are shut (either manually or by a separate sequence system) until the control valve is required for operation.

Three-element control In the three-element system, a controller (item 11 in Fig. 3.23) compares the measured steam flow with the feed flow and modulates the main regulator to keep the two signals in step. Under ideal conditions this would be enough to keep the drum level steady, but it will be apparent that the flows into and out of the boiler and the level of water in the drum form an integrating system, and any small error in the flow measurements will cause the level to rise or fall. This tendency is corrected by the drum level controller (item 8), whose output trims the steam-flow signal upwards or downwards to bring the level to the correct value.

Steam-flow measurement Although steam flow can be measured by conventional means (such as flow nozzles or orifice plates), these represent additional capital expenditure. This can be avoided by making use of the characteristic relationship between the steam flow and the turbine HP stage inlet and outlet pressures. An inferential measurement can be substituted for the steam-flow meter, based on either the turbine nozzle-box pressure or the differential pressure across the turbine HP stage.

(Another reason—though a less significant one—for avoiding the use of conventional flow meters is that, because they produce some degree of unrecovered pressure loss, they also contribute to the overall costs of running the plant and act towards increasing the capital costs since the system auxiliaries have to be larger than they would otherwise be. The contribution will be small in relation to the other pressure drops in the plant, but they are nevertheless real.)

Although the use of inferential measurements is adequate for many purposes, the concept becomes invalid under conditions when some of the steam bypasses the relevant stage of the turbine—as when the unit is fitted with a bypass valve and this opens, or when the machine is run up on the IP stage only. In these cases, various solutions have been effected, such as 'freezing' the HP stage differential signal and switching to an alternative (perhaps the differential pressure across a part of the boiler system) while the flow is diverted.

3.7.2 Feed-Water Control in Once-Through Boilers

Two control approaches will be examined for once-through plant: one as used (largely by European manufacturers) on subcritical boilers, the other as used (mainly by American suppliers) in association with supercritical boilers. One important facet of once-through boilers is common to both types, the fundamental difference from drum-type boilers as far as pre-start and start operations are concerned. This is in part because, as stated in Sec. 2.5, once-through boilers require a precleaning operation with demineralized water before they are started up.

Subcritical once-through boiler feed-water control Under conditions of low flow of once-through boilers (such as during start-up or low-load operation), it is very important to ensure that there is an adequate flow of water through the furnace tubes in order to avoid them fouling. This criterion leads to the development of fairly complex control arrangements.

Figure 3.24 shows the feed-water control system of a typical subcritical boiler in simplified schematic form.[12] The control requirements satisfied by this system are as follows:

88 BOILER CONTROL SYSTEMS

1. At start-up from cold, the feed pumps are initially controlled to maintain the feed pressure at a safe value.
2. For hot start-up, the feed pressure is first held at a value slightly higher than the boiler pressure, so that feed can be provided to the boiler as quickly as possible.
3. As load building starts, flow is controlled by a low-load valve, but as soon as the load approaches the capacity of the start-up valve, control

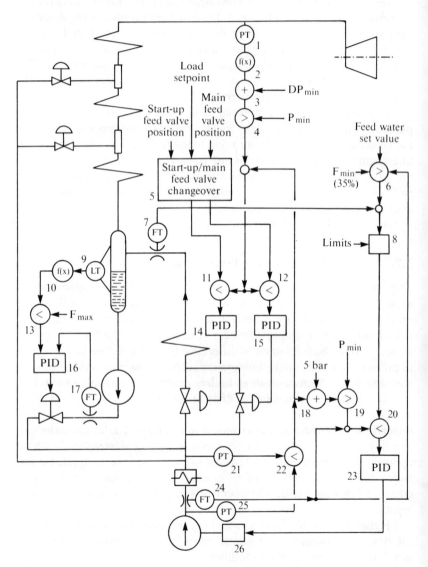

Figure 3.24 Once-through boiler feed-water control (schematic).

is transferred to the main feed regulator and the start-up valve is progressively shut.
4. The safe value for the minimum feed pressure is load dependent, and the system closes the feed regulator to protect the pumps if the pressure drops below this value.
5. The function of maintaining adequate flow through the furnace tubes above a critical value during low-load operation is achieved by recirculating water through the start-up vessel.
6. Changes of recirculation flow are introduced slowly, to prevent the onset of instability due to the positive-feedback effect of changing the cold feed into the start-up vessel.

In the control systems for this type of boiler the 'feed-water set-value' signal which forms the main determinant for the feed pumps, is derived from the unit load demand, but recognizes factors such as the need to keep a minimum flow through the furnace tubes at low loads and the enthalpy gain due to firing changes. In the feed system shown in Fig. 3.24, this signal is directed to a maximum selector to prevent the flow demand falling below a minimum value at low loads.

The output of this selector is compared with the measured flow (item 7 in Fig. 3.24) and the controller (item 23) adjusts the delivery of the feed pump to keep the flow at the desired value. However, if the delivery pressure falls below a minimum value, the output of the pump is automatically incremented to keep the pressure above the safety limit. The pressure signal used for this protection system is derived from the lower of two values: the pressure at the pump discharge or the pressure after the feed heater. Normally, the latter is selected, since it is lower than the discharge pressure due to the pumping loss through the pipework and the heater. However, the second signal is selected at start-up, when the feed check valves are closed.

The selected feed pressure signal is also transmitted to the feed-regulator controllers, and a system of comparators (not shown in Fig. 3.24) is used to transfer the duty of feed-flow control to the valve system at low loads. At high loads, the same arrangement transfers feed-flow control to the pumps.

The load setpoint signal is transmitted by the valve selection block to the start-up valve controller via a minimum selector (item 12). At low loads, this block selects the load demand signal, and the valve (rather than the pump) is therefore modulated to match the demand.

The level of water in the start-up vessel is controlled by modulating a valve in the recirculation line. As any changes in the recirculated flow are also detected by the flow meter in the feed-water control system, the latter compensates for the change by adjusting the feed flow. Again, this is only the basic principle: in practice the actual control system includes several refinements to overcome problems caused by the positive-feedback effects of feed changes. (For example, an increase in cold feed will cause an increase in the amount of steam condensing in the evaporator. This decreases the amount of boiling and reduces the amount of water expelled to the separating vessel, whose level therefore tends to decrease.) These refinements are complex functions in themselves, since the stabilizing delay they apply to the system must not prevent rapid response when a genuine rise in level occurs.

Supercritical once-through boiler feed-water control The procedures involved in starting up one design of supercritical once-through boiler[13] are illustrated in Fig. 3.25. Knowing how this type of unit is started up and loaded is critical to understanding the control strategy that is employed.

Figure 3.25a shows that, during the 'cold clean-up', demineralized water is passed through the evaporative section, the primary superheater and the condenser, with the objective of removing dissolved solids. The hot clean-up is shown in Fig. 3.25b, and it shows that the water path is the same as for the cold clean-up. However, the boiler is now being fired and the temperature of the water gradually rises until steam is flashed off in the pressure-reducing valve A feeding the flash tank. This steam is then passed to the condenser.

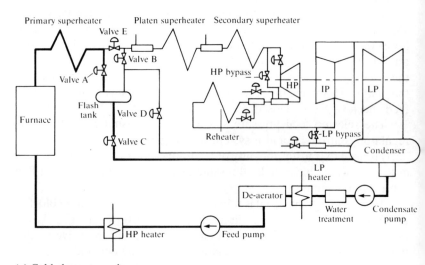

(a) Cold clean-up mode

Figure 3.25(a)–(g) Start-up procedure for a supercritical once-through boiler. (Courtesy of Bailey Controls Co.)

MODULATING CONTROL SYSTEMS **91**

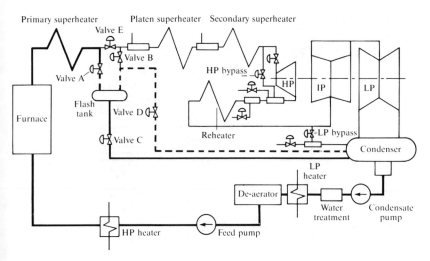

(b) Hot clean-up mode

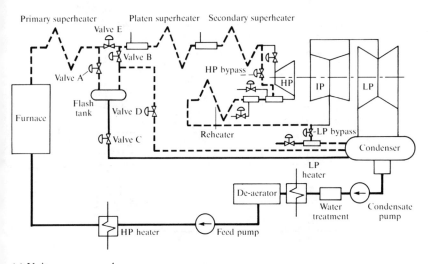

(c) Unit warm-up mode

92 BOILER CONTROL SYSTEMS

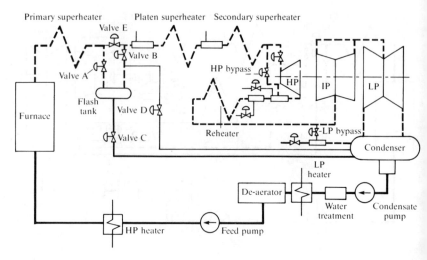

(d) Turbine start-up mode

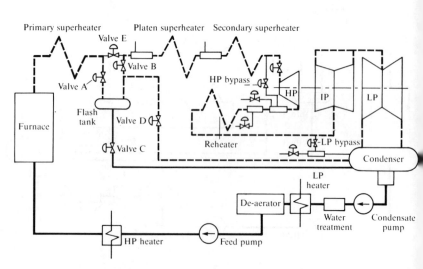

(e) Minimum load mode

MODULATING CONTROL SYSTEMS 93

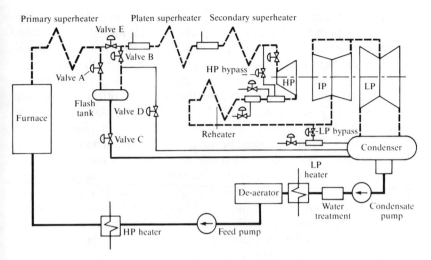

(f) Boiler transition mode

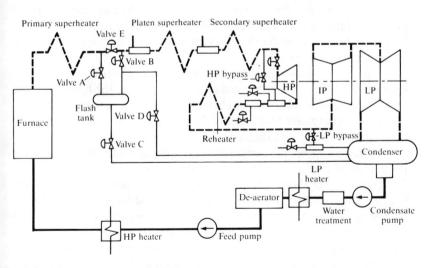

(g) Variable-pressure and full-pressure

At unit warm-up (Fig. 3.25c), valve B has been opened to pass steam through the platen and secondary superheaters. At this stage the turbine is still isolated, and the steam is directed through the HP bypass valve to the reheater before being sent through the LP bypass to the condenser.

Once the steam has reached the required dryness, the HP bypass valve starts to be modulated to control the turbine throttle pressure while the firing rate is slowly increased. During this time valve D is used to control the steam temperature by regulating the division of steam between the valve and the superheaters. Once the steam flow through the HP bypass has reached an adequate rate and the superheater outlet temperature has stabilized, the turbine throttle valve is opened fractionally, causing the turbine to start rolling—the condition shown in Fig. 3.25c. By simultaneously closing the HP bypass and opening the turbine throttle valves, the machine is brought up to synchronous speed and initially loaded. At this stage throttle pressure control is transferred to the turbine valve and the HP bypass gradually closed.

Figure 3.25d shows the system during this transition, with steam flowing both to the turbine and through the bypass. When the bypass is fully closed, shown in Fig. 3.25e, the unit is operating at or below the load that corresponds to the minimum feed-water flow rate to ensure adequate cooling of the furnace tubes and to prevent steam being generated at too low a pressure.

The boiler firing rate is now increased until the load reaches the minimum value, at which point it is held steady. At this time the unit is being operated under the 'turbine-following' régime, with the steam temperature controlled by the pressure being sustained in the flash tank by the turbine inlet valve.

Figure 3.25f shows the boiler being transferred to true once-through operation. Valve A has been gradually shut to isolate the flash tank, and valves C and D (feeding the condenser) are shut when the pressure and temperature have fallen sufficiently. Meanwhile, valve E is modulated to control the primary superheater outlet pressure.

During this time, the turbine throttle valve is gradually moved to the fixed position that it will adopt for quasi variable-pressure operation.

Figure 3.25g shows the system operating under variable-pressure control, with the flash tank now fully isolated and both attemperators in operation. The sprays are used to provide fast control because, although in once-through operation the final steam temperature is dictated by the water flow and the heat input, the response time of this system is long—particularly at the lower loads, when lags of 30 minutes are quite common.

As usual in this régime, the opening of the turbine inlet valve is held at a fixed position and the throttle pressure is allowed to float according to the boiler firing rate—which, together with the feed-water flow, is adjusted to meet the unit load demand. Valve E can now be modulated to meet small, rapid load changes and to keep the primary superheater outlet pressure high enough to maintain supercritical pressure in the furnace circuits.

As the throttle pressure approaches 87.5 per cent load, valve E is opened fully and pressure control is transferred to the turbine inlet valve.

Figure 3.26 (essentially a repeat of Fig. 3.9) shows, in outline form, how this operation is implemented. As the system regulates once-through

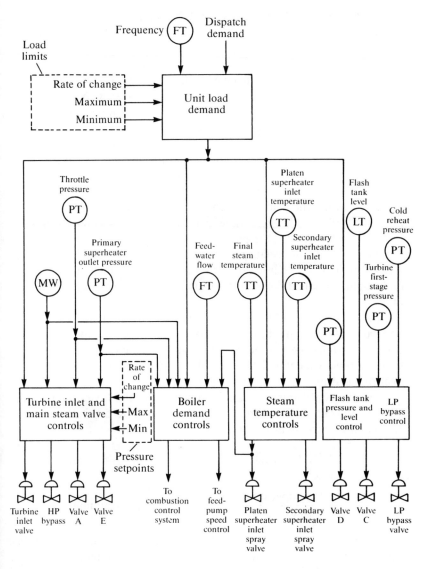

Figure 3.26 Unit control of a plant with a supercritical once-through boiler. (Courtesy of Bailey Controls Co.)

operation, it must coordinate all the major functions, and therefore the unit load demand signal is fed in parallel to the turbine inlet valve and the principal steam valve systems, to the boiler demand (which sets the firing rate and the feed-pump speed), to the steam temperature control loops, and to the flash-tank level and LP bypass subsystems.

3.8 STEAM TEMPERATURE CONTROL

Without the application of any form of control, the temperature of the steam generated by a boiler would follow the 'natural characteristic', with the temperature curving up from a low figure at low loads and tending to flatten out at the highest loads. It is therefore necessary to apply control in order to achieve a reasonably constant steam temperature over the largest possible load range. Since the control system relies for its operation on cooling the steam to bring it to the desired value, it can only work when the temperature achieved by the natural characteristic is above the desired value. In other words, the system cannot function at the lower loads, when the temperature is already below the desired value.

Even within the load range where control is possible, the function is made difficult by the long time constants and the complex transfer functions of this area of the plant.

These comments relate to both the superheated and the reheated steam, though the characteristics of the plant mean that the control philosophies applied to these two often differ significantly from each other.

Also, the philosophy adopted for controlling the steam temperature depends on the design of the plant. Some conventional boilers have non-contact ('drum-type') attemperators and some have 'spray-type' desuperheaters, while fluidized-bed boilers also modulate the flow of solids over the heat-exchanger surfaces to attain the same objective.

The time constants and transfer functions for the steam-temperature systems of a conventional (that is, not fluidized-bed) 500 MW unit[14] are given in the expressions below:

$$\frac{\text{Steam temperature}}{\text{Spray flow}} = \frac{0.72}{(1 + 85P)^3} \, °\text{C/per cent input}$$

$$\frac{\text{Steam temperature}}{\text{Firing rate}} = \frac{5.4}{(1 + 110P)^2} \, °\text{C/per cent input}$$

From these equations it will be possible to gauge the problems of controlling the steam temperature, notably the very long time constants of the system.

3.8.1 Superheated Steam Temperature Control

Fundamentally, the temperature of the final superheated steam is a function of the boiler's firing rate and the steam flow, and of the design of the heating surfaces and the plant generally. In practice, however, the steam temperature of an operating boiler will be affected by the state of cleanliness of the tube banks, and consequently will tend to be higher immediately after soot-blowing has been carried out. However, few systems currently incorporate any provision for recognizing operations such as soot-blowing or the time that has elapsed since the last such operation, although once-through units (by nature of the coordinated unit master) provide a close approach to fulfilling the requirements, leaving the use of attemperators to fine tuning and rapid-response trimming of the temperature.

In comparison, the control systems for the final superheated steam temperature in drum-type boilers rely almost exclusively on attemperators—usually of the spray type. Figure 3.27 shows the elements of a typical system, in which a cascade control system is used to overcome the long time constants of the secondary superheater.

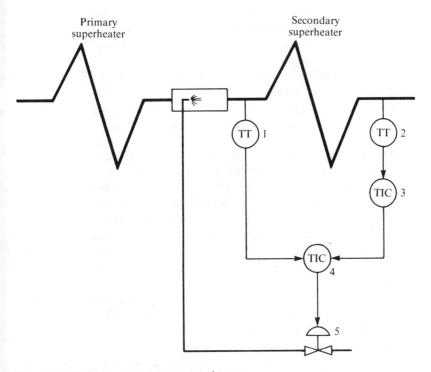

Figure 3.27 Basic steam temperature control system.

In this system, the temperature of the steam leaving the secondary superheater is controlled by a three-term controller. This application is therefore an exception to the rule that, in general, the addition of a third term is rarely possible in power-plant boiler control loops. (Boiler-plant measured variables tend to be very noisy and, since excessive filtering degrades the quality of control, much of the noise will be fed to the controller, where any derivative action will cause the output to become unstable.)

The output of the controller sets the desired value for the slave (item 4 in Fig. 3.27), which compares it with the temperature of the steam immediately after the attemperator. Changes in the steam temperature at this point are quickly detected by the controller and corrected before they can influence the steam temperature at the outlet of the superheater.

Controller saturation effects Although cascade systems of this type are well established, one difficulty—inherent in their design—is poorly understood, and this is the 'saturation' effect (also referred to as 'reset wind-up'). Here it is important to distinguish between 'saturation' in the thermodynamic sense (that is, an indication of the wetness of the steam) and in the control (or electronic, or software) sense. In the latter case, the term refers to a control output that has reached one or other of its limits, so that any further control action *in the same sense* is impossible.

In any controller, there is a limited range over which changes in the input deviation signals result in some alteration in the output signal. Beyond this range the controller output can go no further—its output has reached a limiting condition. The range beyond which such limiting conditions occur is inversely proportional to the gain of the controller: the higher the gain, the more the output changes for a given input change, and therefore the quicker the onset of limiting. (This is one reason why control engineers talk about the controller's 'proportional band'—the band within which output changes are proportional to input changes. The proportional band is defined as being the reciprocal of the gain, so that a controller with a gain of 2 has a proportional band of 50 per cent. In this case an input deviation of 50 per cent produces a maximum output.)

Saturation is a characteristic of any controller that produces an analogue output, whether the controller is of the stand-alone type (pneumatic or electronic) or whether it is computer-based. At first sight, it may appear that pulsed output controllers are not so affected, since they continue to produce output signals under conditions where non-pulsed controllers would have reached a limiting condition. However, if the controller/actuator system is considered as a whole, it will be seen that, since the actuator integrates the controller pulses, the *system* will reach a limiting condition when the actuator has reached the limits of its movement.

One way of thinking about saturation—which will assist in understanding the particular effects of this phenomenon in a boiler steam temperature control system—is to consider the controller as possessing a 'window of visibility', a range of input deviations within which it can respond, and outside of which it cannot.

In a simple loop, this condition is not too important, since the controlled element will start to move 'off its stop' as soon as the input deviation has returned within the proportional band of the controller, but in a cascade loop it has undesirable consequences.

Examination of the steam temperature control system shown in Fig. 3.27 will show that, if the 'slave' (item 4) has a gain of 2, its output will stay within the limits of its possible excursion only while the desired value stays within 50 per cent of its full excursion. But in this case the desired-value signal is actually set by the output of the main controller, so that the remainder of the main controller's output signal range lies effectively outside the 'window of visibility' of the slave. And, the higher the gain of the slave the smaller the 'window' becomes.

Consider the system further, with the main controller gain set at unity and the desired value set to 550°C. If the temperature transmitter is ranged 300–600°C (such suppressed ranges being common in steam temperature control systems of this sort), a measured temperature of 400°C corresponds to a deviation of 150°C, which is 50 per cent of the controller range. In this condition the main controller, with a gain of 1, produces an output of 50 per cent —which is right on the edge of visibility of the slave.

Now, consider what happens when the temperature is, say, 350°C *and rising*. Although the main controller detects the rising signal and reacts correctly, calling for the sprays to come into action and pull back on the rising temperature, its commands are outside the restricted window of the slave controller. This means that the system does not respond at all until the 400°C threshold is reached—by which time it may well be too late.

The long time constants of this part of the plant require rapid response if temperature excursions are to be corrected before they become too great. A saturated controller is powerless to act in these circumstances and it is therefore vital that the control configuration includes an effective method of avoiding the onset of the problem. For example, computer-based systems may adopt software blocks to monitor the main controller output so that when it reaches 100 per cent the 'reset' function is activated (holding the output at 100 per cent) and the controller 'track' function is switched on.

One important consideration in the control of the superheated steam temperature is the need to prevent the temperature at the inlet to the second stage of superheating from falling too near the saturation point.

Apart from the life-reducing effects of thermal shock, such 'chilling' could result in at least partial plugging of the steam circuits with water, resulting in reduced steam flows and the risk of premature tube failure.

Because of the boiler's natural temperature characteristic, spray attemperation is possible only over a restricted range of operation near full load (typically 75–100 per cent of MCR). Moreover, bringing spray attemperation into use at lower loads can result in poor performance because of the exaggerated capacity of the sprays relative to the steam flow. In some installations, however, small-capacity 'start-up' sprays are used in addition to the full-load ones, and in these cases it is necessary to provide smooth transfer of control between these and the main spray valves.

3.8.2 Multistage Superheaters

In boilers with several stages of superheating, spray attemperators are normally provided between the major superheating banks. Figure 3.28 shows a control system for this type of plant, employing cascade systems for each section.

The operation of each individual cascade system is broadly similar to the simple loop described earlier, but the secondary superheater system includes some notable provisions.

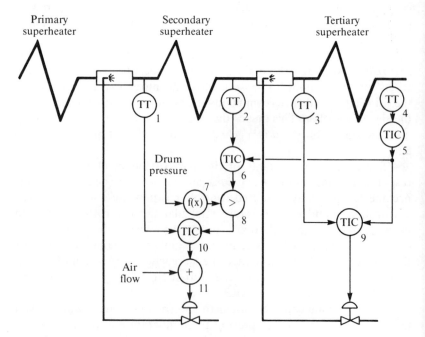

Figure 3.28 Multistage superheater steam temperature control system.

MODULATING CONTROL SYSTEMS   **101**

The first of these is the generation of the desired value for the secondary superheater outlet temperature controller from the outlet of the final steam temperature controller (item 5 in Fig. 3.28). This is because the final temperature controller output signal can be seen to determine the optimum temperature conditions throughout the superheater string.

The second point is the maximum selector block (item 8) interposed between the first-stage main controller (item 6) and its slave (item 10). This block also receives a signal derived from the drum pressure via a function block, which translates the measured pressure into the equivalent saturation temperature, and adds the required safety margin to the resultant. This arrangement prevents the slave from receiving a desired-temperature signal that is too close to the saturation temperature, and therefore prevents the 'chilling' effect referred to earlier.

Finally, the system incorporates an air-flow feed-forward signal in an attempt to optimize response to load changes.

3.8.3 Steam Temperature Control in Fluidized-Bed Boilers

Although conventional spray attemperation is used in fluidized-bed boilers, some designs of such plant enable other measures to be used to provide steam temperature control with improved overall performance of the plant.

For example, in multi-solids fluidized-bed (MSFB) plants with external heat exchangers, it is possible to control the flow of solids to the superheat compartment of the heat exchanger to vary the heat pick-up there. This provides effective control of the plant under steady-state conditions, but the slow response of the solids recycling system requires the use of spray attemperators to meet transient disturbances.

3.8.4 Reheated Steam Temperature Control

The reheater system displays many of the difficulties of the superheat counterpart. For example, the time constants and transfer functions are given by a similar equation:

$$\frac{\text{Steam temperature}}{\text{Firing rate}} = \frac{7.2}{(1 + 110P)^2} \, °\text{C/per cent input}$$

Nevertheless, because of the lower steam pressure in the reheater system, the thermodynamic conditions are quite different from those of the superheater, primarily implying a reduced risk of 'chilling'.

Spraying water into the (lower pressure) reheat circuit has an adverse effect on the cycle efficiency, and therefore it is common to reserve operation of reheat sprays for emergency duty only, with the reheated steam temperature being controlled primarily by adjustment of the hot gas flows over the tubes. In many plants separate gas recycling fans are specifically provided for this purpose.

Moreover, because of the large pressure drop between the feed pump discharge and the reheat circuit, the reheat spray-water supply must be taken from a point before the main boiler-feed pumps. Alternatively, separate pumps may be provided for the reheat sprays, but it is necessary to balance the increased capital and running costs of this scheme against any cost savings that may accrue.

To control the hot gas flows, two separate sets of dampers are sometimes employed, one in the path of the gases flowing over the superheater banks, the other controlling the gases flowing over the reheater banks. To protect the plant against overpressurization, these damper sets are controlled so that one set must be fully open before the other set can begin to close.

Where no gas recycling fan is provided, it is necessary to ensure that an adequate flow of heat is transferred to the reheater at all times. In some instances this is achieved by the provision of separate pressure control dampers in the boiler exhaust ducts, and in such cases it is vital to prevent damage to the furnace walls through overpressurization. This requirement is so vital that mechanical stops are used to prevent closure of the dampers past a certain point.

3.9 THE CONTROL OF COMBINED-CYCLE PLANTS

The control systems for combined-cycle plants must take account of the characteristics of both the cycles. For example, the turn-down range of a

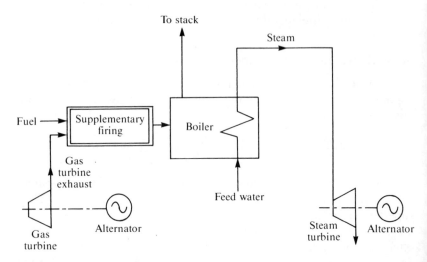

Figure 3.29 Supplementary firing stage in a combined-cycle plant.

gas turbine plant is limited, and in order to achieve wide-range operation of the plant it may be necessary to provide—and modulate—an auxiliary firing system, as shown in Fig. 3.29.

Another variant of the combined-cycle concept is given in Fig. 3.30, which shows the main plant items and their control systems for a combined-cycle plant with a single generator driven by a gas turbine *and* by a steam turbine.

This diagram illustrates the general principle applying to all combined-cycle control systems: that each plant item is controlled by a loop (or loops) similar to the equivalent in a simple (Rankine cycle) plant, at least in essence. The requirement, therefore, is to coordinate the operation of these various subsets of control in the most efficient manner,

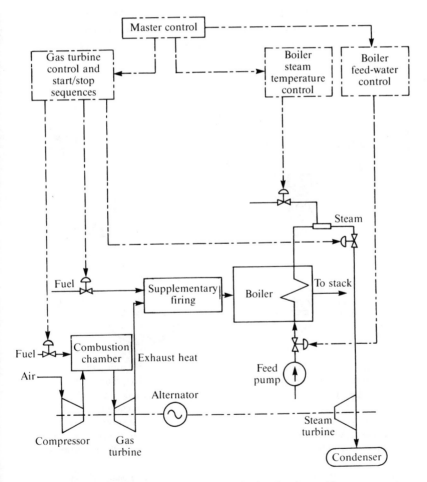

Figure 3.30 The major control loops of a combined-cycle plant with one generator. (Courtesy of Bailey Controls Co.)

104 BOILER CONTROL SYSTEMS

and once again the control philosophy must recognize the strong interactions between the various components of the whole (in particular, the gas turbine and the steam turbine).

These factors dictate that the control of a combined-cycle plant must be achieved through the implementation of a multilevel (hierarchical) control strategy. Figure 3.31 shows one way of achieving this functionality, while Fig. 3.32 shows the practical implementation of this philosophy in a combined-cycle plant. In this implementation, each area of control functions in the way of an 'island of automation', continuously determining the actions to be taken with respect to local parameters—but with the overall strategy (performance targets, etc.) set by conditions and events in associated 'islands' and responding to commands of the functional blocks above it in the hierarchical structure.

This structure assists in meeting another practical characteristic of combined-cycle plants: the fact that it is quite common for the vendors of the main plant items to provide them with 'packaged' control according to their own preferences. In many cases, this results in the situation where

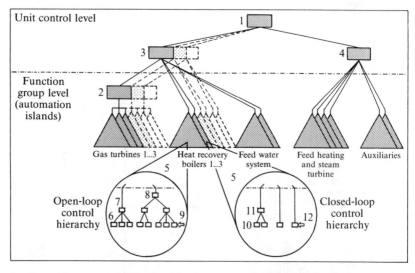

1 Control program at unit control level
2 Gas turbine control program
3 Control programs for heat recovery boiler and feed-water supply system
4 Control program for steam turbine, including feed heating and condensation
5 Control hierarchy of automation islands
6 Drive unit level
7 Drive group level
8 Function group level
9 Protection functions
10 Drive control level
11 Main group control level
12 Control loop override functions

Figure 3.31 Hierarchical control structure of a combined-cycle plant. (Courtesy of ASEA Brown Boveri Power Generation Ltd.)

MODULATING CONTROL SYSTEMS 105

Figure 3.32 Control room console of a combined-cycle plant. (Courtesy of ASEA Brown Boveri Power Generation Ltd.)

the gas turbine is controlled by a system from one manufacturer (and based on one type of microprocessor) while the boiler is fitted with a system from another supplier (and based on a different microprocessor). It is vital to integrate these systems into a cohesive whole that can react correctly and quickly enough as an entity.

Where combined-cycle plants are used solely for the generation of electrical power, it is usual to operate the boiler 'flat out' (that is, to convert into steam the maximum possible amount of heat in the gas turbine exhaust). The supervisory function in the control hierarchy is then used to adjust the gas turbine output so that the combined (steam and gas turbine) plant meets the overall electrical load requirement.

3.10 THE CONTROL OF CO-GENERATION PLANTS

The control systems for co-generation plant are more complex than those of ordinary power generating units since they must recognize the different demands of both the electrical and the thermal loads and optimize the usage of the plant to serve both requirements. Since many of today's co-generation plants also embody combined-cycle technology, the control techniques described in the previous section often need to be incorporated into the overall philosophy of the scheme.

106 BOILER CONTROL SYSTEMS

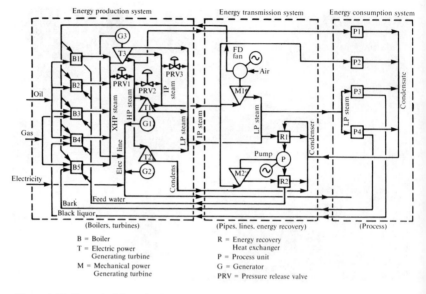

Figure 3.33 Control philosophy for a pulp and paper plant. (Courtesy of Bailey Controls Co.)

The actual control arrangements will be very much determined by the plant configuration and the functional requirements of the entire installation. Typical factors could be whether the installation is merely using surplus steam for district heating or whether it is providing steam for a process that is a prime consumer.

Figure 3.33 shows the control philosophy for an industrial cogeneration facility providing steam at two pressures to a process while at the same time generating electricity (some of which is exported). The requirements here are:

- to generate the optimum level of electrical power;
- to provide sufficient process steam at the two pressures;
- to minimize the quantity of steam passing to the condenser;
- to operate the turbo-generators within design constraints;
- to minimize pressure reduction operations during transients.

One implementation of such a system[15] uses an advanced control strategy that automatically modulates the speed/load reference for each of the two turbo-generators to maintain at a calculated setpoint the value of the electrical power purchased from the utility. This purchased-power setpoint represents the power consumption which, if maintained at that steady state, would result in the power demand exactly reaching the maximum limit entered into the controller (Fig. 3.34). Separate target

MODULATING CONTROL SYSTEMS 107

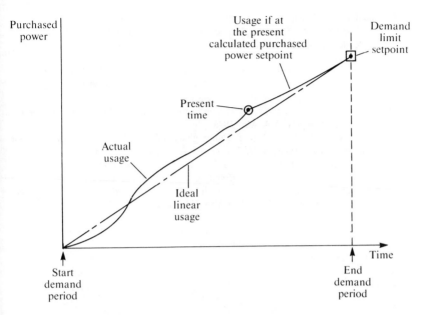

Figure 3.34 Control of a co-generation plant to meet target electrical load. (Courtesy of Bailey Controls Co.)

demand limits can be set for different times, e.g. for on-peak and off-peak operation.

Fortunately, since simply controlling the purchased power to a set level would result in excessive modulation of the turbine speed and load, and fluctuations in the supply of steam to the process, it is generally necessary to control only the total electrical power purchased during the demand period.

The philosophy uses an interesting variety of adaptive control, since the parameters (gains and deadbands) are automatically adjusted over the demand period, with wider deadbands and lower gains being used at the beginning of the period. Towards the end of the period the 'decrease' deadband is opened up while the 'decrease' gain is reduced, so that the on-site generation is not reduced excessively, only to be reinstated at the beginning of the next demand period.

The system also incorporates provisions to avoid incurring penalty charges during the transition from peak to off-peak periods, while at the same time avoiding disturbances to the supply of process steam.

NOTES

[1] The term 'device' is used here, which is relevant to a hard-wired system. In practice, all such control functions may be served by software in a computer system.

[2] F. Hagedorn and G. Klefenz, *H & B Power Station Control—Part I: Unit Load Control*, Monograph 02/66-3587 EN, Schoppe and Faiser GmbH, Germany.

[3] A. H. Rudd, Variable Pressure Boiler Operation, Canadian Electric Association, Ottawa, 1972.

[4] Described in greater depth in publications E86060-T6074-A201-A1-7600, E86060-T6074-A111-A1-7600, E86060-T6074-A321-A1-7600 and E86060-T6074-A221-A1-7600, from Siemens AG, ZVW 85, Fürth-Bislohe, Germany.

[5] H. D. Ege, Jr, 'Control Systems and the Combustion Process', *CIBO Seminar on Control Developments*, Arlington, VA, May 1988.

[6] D. Anson, W. H. N. Clarke, A. T. S. Cunningham and P. Todd, 'Carbon monoxide as a combustion control parameter', *J. Inst. Fuel*, vol. xliv, no. 363, April 1971.

[7] F. H. Holderness and J. J. MacFarlane, 'Continuous flow instrumentation analysis of flame gases, *Seventh Symposium on Combustion*, Butterworth, London, 1958.

[8] D. V. Eskenazi, *et al.*, 'On-line monitoring of unburned carbon', *EPRI Heat Rate Improvement Conference*, EPRI, Palo Alto, CA, September 1989.

[9] See note 1 in Chapter 2.

[10] The Battelle design of MSFB boiler, as implemented by Foster-Wheeler Power Products at Kerry Milk Products (Ireland) and Slough Estates (UK).

[11] 'Hydrastep' is a trademark of Schlumberger Industries.

[12] The diagram and the accompanying description are both necessarily simplified. A more detailed explanation will be found in the monograph, *Feedwater Supply for a Once-through Steam Generator with a Circulating Unit* (order no. E86060-T6074-A151-A1-7600) available from Siemens AG, ZVW 85, Fürth-Bislohe, Germany).

[13] A Babcock and Wilcox 'Universal' boiler. Systems and description courtesy Bailey Controls Co., Wickliffe, OH 44092, USA.

[14] Adapted from M. F. Binney, 'General aspects of control system design in power stations,' *Conference on Automatic Control in Electricity Supply*, IEE, Manchester,' March 1966.

[15] 'Advanced Cogeneration Steam Turbine Control'; Bailey Controls Co., USA; Application Guide AG-0000-051-01.

CHAPTER
FOUR
ANCILLARY SYSTEMS

In addition to the main systems described in the preceding chapter, boilers are provided with many additional control loops. In general, these are minor systems, typically handled by self-acting controllers, and require very little explanation. For this reason, and because of the number and variety of such systems employed on power-plant installations around the world, these systems are not described in this section.

One exception, however, is the major and complex issue of the turbine bypass system. Properly a part of neither the boiler nor the turbine, yet integral to the operation of both, turbine bypasses are nowadays essential to many installations.

4.1 TURBINE BYPASS CONTROL

Practically, many different arrangements of turbine bypass are encountered on power plants. In the simpler examples, a single valve bypasses the HP stage of the machine and another the LP stage. In more complex installations, an arrangement of small valves is used to bypass steam from various parts of the boiler to the hot reheat line or to the condenser. These valves may be ranged at 3–5 per cent of the boiler MCR, and are used to assist with cold starting or hot restarting of the unit. (In one example, use of small bypasses of this type combined with modified sliding-pressure control is reported to permit hot restart of a shut-down 500 MW unit with a coal-fired drum boiler in just over 90 minutes.[1])

The control arrangements for these systems range from simple manual facilities to complex arrangements that are fully integrated with the main plant start-up sequences and on-line modulating control systems. The number and variety of such systems preclude description in a general overview of control systems such as this.

With large bypasses, however, although the control systems must include provision for several interlocks with the other plant operations and controls (making them very dependent on the details of the particular plant configuration), the basic philosophy of control is well defined and fairly widespread.

4.2 HP BYPASS

As shown in Fig. 4.1, the HP bypass valve is under the command of a controller, which modulates it to keep the main steam pressure at the desired value during start-up and shut-down. A feed-forward signal is included in the control system, opening the bypass in the event of a sudden reduction in load. In order to give adequate response in the latter condition, some manufacturers provide a separate high-speed actuator for use with the loss-of-load loop.

When the unit has been loaded, and under normal running conditions, the pressure control loop setpoint is arranged to 'shadow' the normal master pressure setpoint so that the bypass is brought into action if the pressure rises abnormally. In this respect the bypass functions as a relief valve, reducing the frequency of operation of the safety valves. (Safety valves tend to be noisy in operation, and anything that reduces the calls for their operation is desirable, especially near residential areas.) Extrapolating from this relief-valve function, some countries allow bypasses to assume the rôle of a safety valve (at least on once-through boilers) without the need for separate, dedicated safety valves. (Obviously, in such cases the bypass must be rated at 100 per cent MCR!) When the bypass is used as a safety valve, the design of the control system assumes a particular importance, since no conceivable failure can be allowed to prevent the opening of the valve when needed.

Figure 4.1 shows the non-return valve that is required in the cold reheat line from the HP stage, to prevent the turbine from suffering a reverse flow of steam via the bypass valve. (In practice, such a valve is sometimes provided on units without bypass, with the purpose of preventing water from being induced into the HP stage from any leakage from the reheat sprays during and after a plant trip.)

Temperature control is required, in order to cool the bypassed steam to the level of the normal HP stage exhaust. A separate control loop is provided for this purpose, regulating the temperature of the steam leaving the valve by adjusting the admission of cooling water. The water is usually injected near (or directly into) the valve seat, so as to give good control even when the valve has suddenly opened, and to assure adequate

ANCILLARY SYSTEMS 111

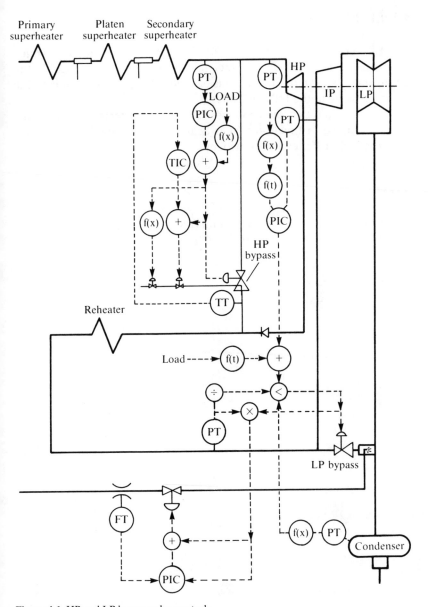

Figure 4.1 HP and LP bypass valve controls.

mixing of steam and water. The sensing point for the measured temperature is located in the line immediately downstream of the bypass valve.

It should be noted that the control systems on the larger units are complicated by the fact that it is usual to have at least two HP bypass valves operating in parallel.

4.3 LP BYPASS

The LP bypass is also provided with a pressure control system, which maintains the steam at the LP stage inlet at the desired pressure. In the arrangement shown in Fig. 4.1, this desired value is set according to the steam pressure at the HP stage inlet.

It is essential that the condenser should be protected against being overpressurized, a function that is achieved by the minimum selector after the main LP pressure controller.

NOTE

[1] H. H. Nelken, W. P. Gorzegno and C. J. Brigida, 'Co-ordinating powerplant components for cycling service,' *American Power Conference*, April 1979.

CHAPTER
FIVE
SEQUENCES AND INTERLOCKS

The foregoing chapters of this book have been concerned with the modulating control systems of the boiler plant. In practice, these are used in conjunction with interlock, inter-tripping and sequenced start-up systems whose complexity and importance match—and possibly exceed—those of the modulating loops.

The interlock, inter-trip and sequence systems are essentially 'on/off' in operation and do not rely on precise modulation of valves or dampers and the like. They are therefore often referred to as the 'binary' systems of the power plant.

The interlock system is the 'bottom line'—protecting the plant against any conceivable malfunction of the control system or the plant itself, and guarding against any errors made by the human operator. The safety of the plant and its operators therefore depends on the systems, and it is vital that they are totally reliable. This is why they must be separated from the system that performs the modulating and start-up duties and, in fact, why they are often entrusted only to hard-wired assemblies that employ the minimum number of components of any type. It should be noted however that, with the emergence of rigid safety standards for programmable electronic systems, many authorities are now prepared to accept the use of programmable logic control (PLC) systems for safety interlock purposes. In such cases, however, it is absolutely vital that the systems, hardware and software *do* actually comply with all the relevant legislation.

In many ways tightly bound up with the interlocks, the inter-tripping system monitors the plant and trips one or more related auxiliaries if a plant item trips.

The sequence functions are provided to assist the human operators by automating the start-up or shut-down of the plant (or a part of it). The sequences are *not* safety functions, and therefore can be executed within the same computer system that carries out the modulating operations. To avoid duplication of functionality, it is also acceptable for them to rely, to some extent, on functions that are executed by the interlocks.

The interlock and sequence systems differ from the modulating loops in one important respect: they relate to the operation of the plant auxiliaries—either as single items or as interacting combinations—and not to the dynamics of the plant. Thus, whereas in the design of a feedwater control system consideration must be given to the flow of water into the boiler and the steam out of it, plus the effects of these on the drum level, the design of a feed-pump interlock or sequence system will be largely related to the pumps themselves. (That is: Can they be safely started at this time? Are the leak-offs open? And so on.) Even the burner management system, which demands consideration of what is happening in the furnace and in the fuel and air supplies, can be designed without too much consideration of the plant dynamics.

This leads to another important distinction: between the *analysis* of the two classes of system. It is quite practicable to talk in general terms about 'three-element feed-water control' or 'cross-limited combustion control', omitting the consideration of details (such as the number of feed valves or pumps in the former case, and numbers of fuel valves in the latter), knowing that it is valid to apply the principles embodied in these concepts to a particular plant. However, it is not possible to discuss interlock and sequence systems other than in a very general way, since the design of these systems is very specific to the number and type of auxiliaries fitted to the installation under consideration. Therefore, only general guidance on some aspects of these systems can be provided here.

5.1 INTERLOCK SYSTEMS

Since the objective of the interlock system is to provide a 'catch all' safety function, the overriding consideration is to prevent the occurrence of any situation that could endanger plant or personnel. Whether or not the use of programmable logic systems is accepted by the relevant authority for this critical function, the design of the systems should lead to the simplest possible configuration.

Once again, there is a vital difference between an interlock system and the modulating loops. In the former, *the system can override the human being*. (In fact, the interlock system will often be able to override the sequence or modulating systems as well, but that fact is not quite so significant as the concept of a machine being given greater power than the human operator!) The reason for this is the fact that the interlocks must guard against incorrect actions (accidental or deliberate) made by a human being. The implication is that, at the design stage, the functional definition of the interlock system must be such that absolutely correct operation of the plant is ensured under all conceivable operational conditions.

Wherever possible, interlocks should tie together switchgear without the use of any active (mechanical or electronic) components. Usually, however, some logic functions will be required and, whereas many authorities demand that these should be relay-based, others are quite willing to accept the use of programmable logic controllers. In these cases, the PLC system design (hardware and software) must comply with the highest standards of safety and reliability.

Although the requirements for protection inevitably lead to the need for providing direct 'hard-wired' connections between the initiating devices and the logic equipment, and between this and the switchgear, it is possible to combine these with information carried on a data highway, to reinforce the indications and to provide additional alarm and control functions.

The interlock systems can be grouped into two broad areas:

Mechanical protection   Where plant drives are protected from damage.
Boiler protection   Protecting the boiler from damage or stress, and personnel from danger.

Usually, the boiler protection logic can itself be divided into seven groups:

1. Master fuel trip logic.
2. Support energy logic (electric, pneumatic and hydraulic supplies).
3. Loss of fuel logic.
4. Turbine trip (and, where relevant, bypass abnormal) logic.
5. Excess firing logic.
6. 'Furnace pressure protection' logic.
7. Environmental protection logic.

Although the interlock systems must provide protection at all times, they must also be designed in such a way that they do not cause spurious shut-downs. For this reason, two-out-of-three 'voting' arrangements are sometimes demanded for critical initiating measurements. With these systems, three independent but identical measurement devices are provided, with the logic system continuously comparing the status of the signals from them. If the status of one measurement is different from the others, an alarm is raised, but a trip is not directly initiated. If two or more agree on an alarm status, the trip *is* initiated.

5.2 SEQUENCE SYSTEMS

The start-up and shut-down operations of the plant, as well as certain intermediate operations (such as transferring control from one item to another), are performed by sequence systems. These systems start up or shut down each item (or group of items) with respect to the safe operational criteria of the item (or group) as an individual entity and as an integrated whole.

Because of the complexity of the equipment being controlled, there are significant risks of a sequence 'hanging' because of, say, a sticky limit switch on an actuator, or a valve that does not open quite far enough. Because of this, and in view of the need to start up and shut down the plant as quickly as possible, sequence systems are generally limited to performing only a small subset of operations. (In other words, it is generally considered inadvisable to have the whole power plant start up at the touch of a single button.)

Furthermore, many authorities discourage the use of general override facilities (which can be extremely dangerous) and instead demand that any failure should be accompanied by a clear indication of the stage reached at the time, so that the operator can complete the relevant sequence manually.

The systems should include provision for warning that a particular sequence has been suspended by a fault. In general, this can be detected by timers that allow a large enough interval for the relevant actions to be completed and raise an alarm if the sequence has not been completed—or has not reached a defined condition—within the allocated time. Such checks need to be applied judiciously, however, since an overabundance of them can cause unacceptable delays.

Because many plant items on a power station are very large, and therefore consume a lot of power, one special requirement of the systems is that such items are interlocked so that only one can be started at a time. This prevents the supply voltage being reduced unduly by simultaneous starting of large motors.

5.2.1 A Typical Sequence—Start-up of an FD Fan

Figure 5.1 shows a start-up sequence for the forced-draught fan of a 125 MW gas- and oil-fired boiler. Certain permissives are checked before the system will inform the operator that the sequence is available for use. Most of these are concerned with the protection of the fan itself (such as the availability of cooling water), but the sequence cannot be started unless at least one ID fan is running. This precaution guards against overpressurization of the furnace.

SEQUENCES AND INTERLOCKS **117**

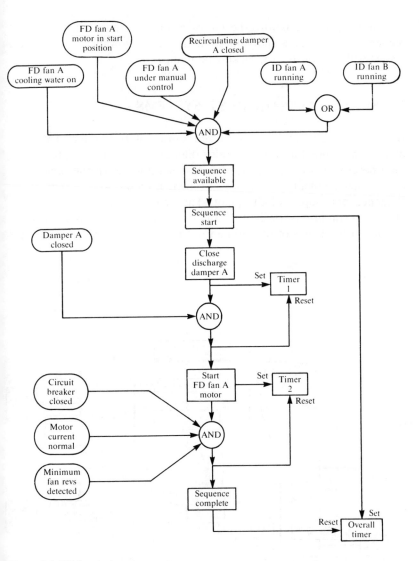

Figure 5.1 FD fan start-up sequence.

As soon as the sequence is initiated by the operator, the fan discharge damper is closed, after which the fan motor is started. As soon as the fan has reached a certain speed, with no abnormal current flowing to the motor, the 'sequence complete' signal is initiated.

Timers monitor the sequence as a whole, and certain subsections of it, warning if an action is not completed within a reasonable interval.

This example is a very simple one. More complex systems have to cover for specific features such as the availability of lubricating oil and jacking oil pressure before the motor is started (though these should recognize that after a short-duration trip it is allowable to attempt a restart without starting the jacking oil pump).

5.3 BURNER MANAGEMENT SYSTEMS

Burner management is a specialized subset of the boiler protection philosophy—albeit a rather complex one. However, since it also assists the operator in starting up, loading and shutting down the boiler, the burner management system may also be considered to be a part of the sequence philosophy of the plant. The systems have to provide safe operation conditions for light-up and operation of the burners as single entities or as groups of related items. They must ensure that no burner

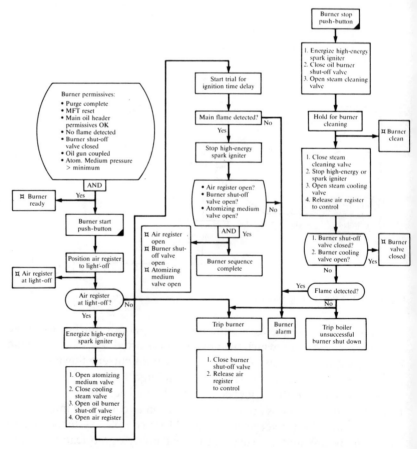

Figure 5.2 Oil-burner management logic summary. (Courtesy of Bailey Controls Co.)

SEQUENCES AND INTERLOCKS 119

can be ignited unless the furnace conditions are safe to do so (this necessitates, for example, a complete furnace purge if the burner is the first one to light) and that failure to establish ignition within a specified time shuts off the fuel supply to prevent explosive ignition at a late stage.

The systems usually employ piezoelectric igniters for pilot gas (or light oil) burners, which in turn ignite the main flame. The logic systems depend on optical flame scanners to detect the pilot and main flames, and these may respond in the visible, infrared or ultraviolet parts of the spectrum. The selection, installation and maintenance of these scanners is very important to the safety and reliability of the process, since flames can be partially or completely obscured by smoke, unburned fuel particles, dust or dirt on the scanner optical parts. Adjacent-burner discrimination should be high, to avoid mistaking a nearby flame for the one being monitored, and the system should also be able to distinguish between the flame and the heat radiation from refractory.

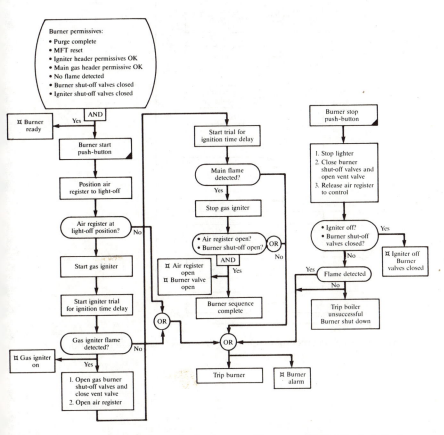

Figure 5.3 Gas-burner management logic summary. (Courtesy of Bailey Controls Co.)

120 BOILER CONTROL SYSTEMS

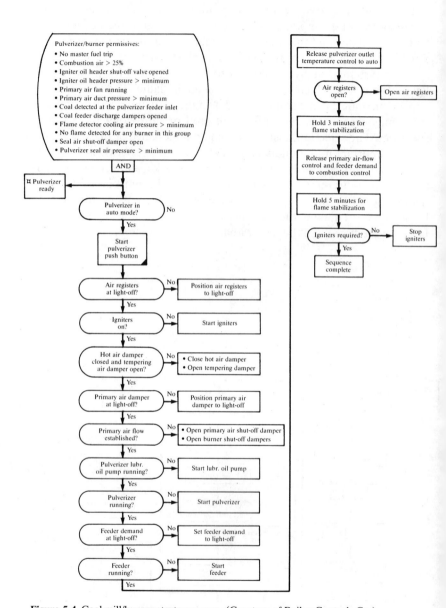

Figure 5.4 Coal mill/burner start sequence. (Courtesy of Bailey Controls Co.)

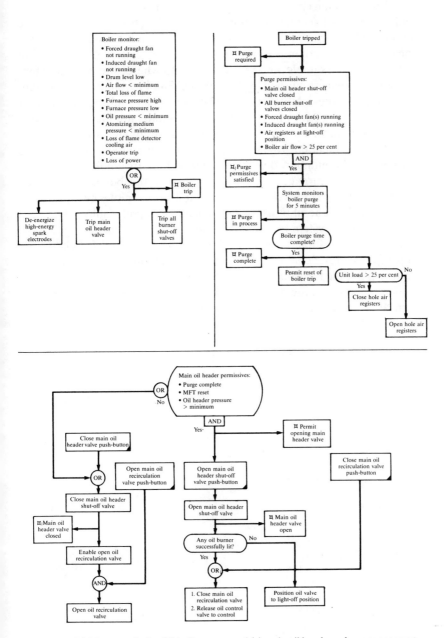

Figure 5.5 (a) Master fuel trip, (b) boiler purge and (c) main oil header valve management. (Courtesy of Bailey Controls Co.)

Burner management is an area where some complexity is inherent, and it therefore becomes necessary to use solid-state or relay-based logic systems. Once again, the safety of the plant depends on the reliability of these systems, and in various countries strict codes of practice are enforced. For example, in the USA and elsewhere, the relevant standards have been established by the National Fire Prevention Association. Variations of these NFPA codes relate to single- and multiple-burner installations, gas-, oil- or coal-burning plants and multiple-fuel boilers.

The design of burner management system applied to a particular plant will be dictated by the fuel, fuels or combination of fuels being burned, by the number, design and type of burners employed, by their location in the furnace and by the design of the boiler as a whole. Considerable variations therefore exist, and any examples given here can only show the general basis for the management philosophy. Figure 5.2 shows a typical system for a multiple-burner oil-fired boiler, Fig. 5.3 the system for a multiple-burner gas-fired plant, while Fig. 5.4 shows the start-up logic for a coal mill and burner combination. These systems are shown in outline form only, and are of course attended by others for main fuel trip, boiler purge and main fuel header valve-management logic, typical examples of these being given in Fig. 5.5.

CHAPTER
SIX
EQUIPMENT CONSIDERATIONS

Almost every new advance in electronic or computer technology (as applied to control systems) is likely to be greeted with some dismay by power-plant engineers. The reasons are as follows:

1. The advent of new control technology makes it increasingly difficult to find maintenance staff who are experienced in the older version, or who are prepared to work with it.
2. As the newer semiconductor devices appear, their predecessors first become obsolescent and eventually obsolete, making it difficult for control systems manufacturers to keep stocks of parts to supply spares for the older systems. (After all, control systems suppliers are not large-scale consumers of the devices and they cannot therefore enforce support for the older models from component suppliers.) Also, it is not generally economic for the manufacturers to design replacement sub-assemblies for the old systems, using newer components. Spare parts therefore become difficult—if not impossible—to find.
3. In spite of this, the control systems continue to be critical to the safe and efficient operation of the mechanical and electrical plant, which, in spite of advancing years, continues to operate satisfactorily.
4. The plant owners and/or operators are therefore faced with the need to refurbish or replace control systems within a timescale that is short in comparison with the life of the main plant.
5. Apart from presenting difficulties of refurbishing control systems on plant that must continue to operate with minimum disruption, the advent of new systems necessitates the retraining of maintenance staff and operators.

In spite of these factors, control technology continues to advance—and there seems to be little prospect of the pace slackening. The solution is to accept the fact and to design the whole installation so that changes can be incorporated with the minimum of disruption.

The following paragraphs are an attempt to delineate some of the fundamental requirements for power-plant boiler control hardware and software. Because of the importance of one subject, manual back-up, detailed consideration is first of all given to the manual control facilities needed for the safe operation of boiler plant.

6.1 MANUAL OPERATION

Throughout the earlier passages of this book, references have been made to the need for ensuring safety and security, and fundamental to this requirement is the provision of back-up manual control facilities, which enable the human operator to control the plant when the control systems are not available, or when they fail.

The manual control facilities provided on some installations, while being quite adequate for many process applications, fail to recognize the hazardous operations and the multivariable, interactive nature of boiler control loops. The weaknesses of such arrangements are not easily apparent, and in many cases (due mainly to the inherent reliability of modern large-scale integrated circuits and other component-level devices) the systems incorporating them can function safely for years.

This does not mean that they are *safe*. The dangers inherent in their design are merely lying dormant.

Unfortunately, some system vendors either do not recognize the risks or dismiss them without realizing the consequences, and plant people cannot be expected to be sufficiently aware of the finer nuances of electronic system design to appreciate the risks.

A somewhat detailed analysis of the problem is therefore justified.

Figure 6.1a shows a simple circuit of a computer driving the input stage of an actuator system (such as an electropneumatic converter). The output signal from the computer takes the form of a current source and, since the signal is generated by the digital computer, the source is driven by a digital-to-analogue converter (DAC). Several computer systems rely for manual operation on the DAC receiving commands that are operator-controlled but still generated by the same computer that provides the automatic control function.

Such a system has several vulnerabilities, two of which are as follows:

- a failure within the computer system will disable manual control as well as automatic operation;
- a fault in the DAC will result in the output taking up an arbitrary value that cannot be corrected by switching to manual control.

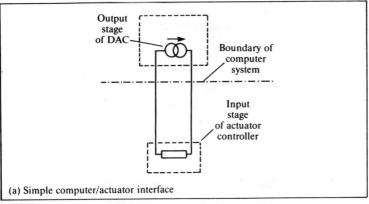

(a) Simple computer/actuator interface

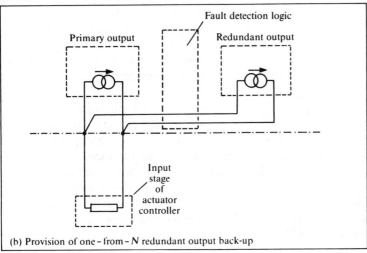

(b) Provision of one-from-N redundant output back-up

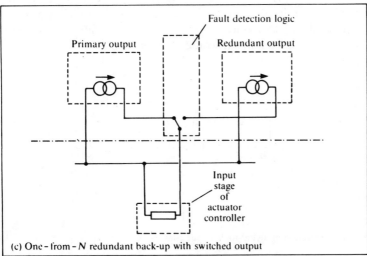

(c) One-from-N redundant back-up with switched output

Figure 6.1 Computer-system analogue output and back-up configurations.

The second of these conditions requires further consideration.

The DAC works by generating an analogue output that is determined by the status of the input data 'bits'. For example, in the eight-bit DAC represented by Fig. 6.2, the value of the output is determined by the bit pattern of the eight input terminals. No bits set (that is, no input at a logic 1 level) produces an output of 0 per cent (which corresponds with a final output of 4 mA in a 4–20 mA range). The four least-significant bits (LSB) being set (to logic 1) corresponds with a digital 15 (1 + 2 + 4 + 8), and 15/256 corresponds with around 5.86 per cent or 4.7 mA. The most-significant bit (MSB) being set corresponds to 128 or virtually 12 mA.

Now, the DAC actually comprises several transistors within a single integrated circuit, and one possible failure mode of this arrangement is that one of the transistors can 'latch-up'—that is, it locks onto the ON state, and stays there. If this happens to be the most significant bit, the system will produce an output of 12 mA. This bit is permanently set ON, and *any attempt to change the output by altering the input bit pattern will produce an output that is 12 mA higher than the one actually desired.*

If the DAC is common to the manual and auto circuits, no amount of adjustment by the operator will change this situation, and in such a case the only remedy is to trip the unit or to shut an isolating valve or damper in an attempt to salvage something from the situation.

Such actions will have to be performed by the operator, who at that time will probably be under some stress. The operator must nevertheless quickly understand what is happening and carry out the correct action (or chain of actions). In some cases the control facilities necessary to correct the situation will not be available on the control desk, and it will then be necessary for somebody to reach the relevant item of plant quickly enough to avert disaster. Given the large physical size of a power plant and the tendency to reduce manning levels to an absolute minimum, success in such an endeavour would be improbable.

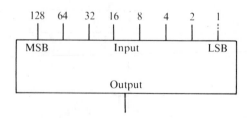

Figure 6.2 Operation of digital-to-analogue converter.

Even where sufficient facilities are available at the control desk to avoid the occurrence of a serious incident, the fact is that at such a time the operator will be trying to cope with several things happening at once. In these circumstances, doing the right thing in an adequately short time would be quite difficult.

Various bizarre solutions have been offered in instances where the existence of this problem has been realized, but one of the more hygienic arrangements is shown in Fig. 6.1b. Here, the analogue output is monitored and, if it assumes a value that the computer does not recognize (corresponding to the 'stray bit' situation described earlier), a 'redundant output' is brought into play. This is essentially a spare DAC and an output driver via which the operator can control the actuator, and one such output is normally provided for a small number of so-called primary outputs (typically seven—hence this becomes a 1-from-7 redundant output).

This is better, but it will be seen that the redundant output is switched into play *in parallel with the (faulty) primary input.* Since the redundant output is added to the primary one and cannot subtract from it, the basic weakness of the simple system is still present in this configuration.

Figure 6.1c shows a similar arrangement, but where the faulty output is actually switched out of the circuit and replaced by the redundant signal. This overcomes the problem caused by the stray bit, but it requires considerable complexity:

- The redundant output must assume the same value as the primary output held before it failed.
- The fault detection logic must be fairly complex, and it must terminate with a relay contact, which itself adds to the cost and increases the risk of failure.

This is why many authorities specify a manual drive system, which is independent for each actuator, is totally autonomous of the computer system and can be brought into play by the operation of a simple switch.

Better still, some authorities demand that output commands from the computer should take the form of stepper-motor drives. The simplest of these produces (effectively) 'raise/lower' commands, in the absence of which the actuator remains in its last commanded position. Manual control is achieved by simply disconnecting the output pulses from the computer and substituting the commands from operator-controlled raise/lower buttons.

Such simple stepper-motor drives suffer from having an inherent deadband, which limits the positioning accuracy of the system, and a better approach uses true multiphase drive commands. With the simplest of these, a two-phase train of pulses is produced: if phase A leads phase

B, the stepper motor rotates in one direction, whereas if phase B leads phase A, rotation is in the opposite direction. Since the phase difference between the two pulse trains can, if necessary, be zero, such a system has no deadband and therefore offers a high level of positioning accuracy.

Manual control with this configuration is slightly more complicated, since some electronic devices are required to generate the two-phase output that is needed, and there is a risk that failure of one of the two phase signals (or the breakage of a single connection between the driver and the actuator) could cause the actuator to move unexpectedly. Nevertheless, experience has shown that the safety, integrity and accuracy of such systems are of the highest order.

6.2 ENVIRONMENTAL CONSIDERATIONS

The environment in which power station electronic systems must operate is very severe from several different standpoints. These can be, and frequently are, defined in precise detail by the authorities in terms of voltage transients, temperature variations and so on, but the reasons *why* these conditions arise are rarely explained. Perhaps the reasons should be obvious, but it is still worth considering them here.

6.2.1 Power Supplies

Astonishing as it may seem in a power generating plant, the constant availability of electrical power for the control systems cannot be guaranteed unless provision has been made for it at an early stage in the design of the installation, and even then the quality of the supply is not ideal. Figure 6.3 shows the type of electrical supply system that may be encountered in a power plant in Europe, and similar arrangements (albeit with different voltage levels) will be met elsewhere. The instrumentation supply (shown in the chain-dotted box to the lower right of the figure) is provided with main and standby supplies, and also power from an inverter system (which is itself provided with a bypass to allow for inverter outages).

More complex arrangements are sometimes provided, perhaps employing standby diesel generators.

Provided reasonable arrangements are made, the supply will be relatively secure, and it is therefore referred to as an 'uninterruptable power supply'. Even in these situations, however, consideration should still be given to the security of the plant when the power *does* fail, even for short periods—and this becomes even more important where less secure supply arrangements are provided.

Being at the generating end of the power system, the plant electrical

EQUIPMENT CONSIDERATIONS **129**

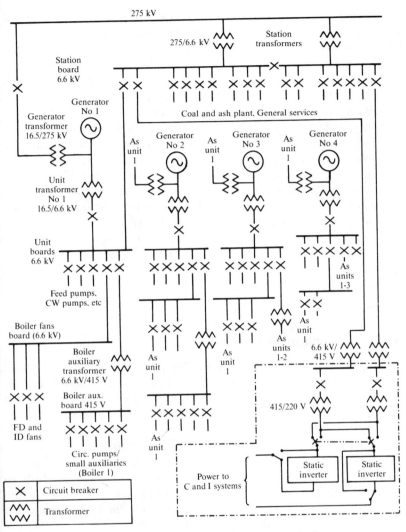

Figure 6.3 Part of a power-plant electrical distribution system (typical).

supplies are subject to severe transients and interruptions: in simple terms, if 600 MW of energy (or more) is suddenly dumped due to a generator fault, the supplies in the vicinity are likely to experience some transient disturbances!

The transient voltage 'spikes' that may arise on a given power line will usually be specified by the relevant authorities and will be related to the nominal voltage of the relevant supply, but a requirement for equipment to survive a 1200 V 10 μs transient on a 115 V line is not unusual.

130 BOILER CONTROL SYSTEMS

At the other extreme, owing to the way in which the systems are connected to local supplies, generators and the external network, partial or complete losses of power will be experienced, typically while switchgear is operating. Some authorities specify different values for 'brown-out' and 'black-out' conditions (the former being a temporary reduction in the supply voltage, the latter a complete loss of power). Certainly, modulating and protection equipment should continue to operate satisfactorily, and protect the plant, under the specified conditions.

Following a complete and extended loss of power, it should clearly be possible to restart the plant with the largest possible degree of help from the electronic systems, and in the shortest possible time. Control configurations and parameters should therefore be retained, and this implies some form of battery back-up (for the memories at least). It is not advisable to use so-called 'station batteries' to power electronic systems—the voltages produced by such supplies are very variable and, as very strange electrical loads are sometimes connected to the batteries, and highly inductive loads suddenly switched on or off, extreme transients will be met.

6.2.2 Electromagnetic Compatibility

The operation of switchgear controlling power to motors consuming several megawatts, and the transients resulting from generator faults, are just two of the reasons why the levels of interference experienced in power plant are so high—and why the electronic systems must be immune to them.

Another factor to be considered is the extensive use of portable communications systems ('walkie-talkies') on site—typically during construction and commissioning. The situation often arises in which an engineer investigating a problem will open the doors of a cubicle and press the 'transmit' key of his transceiver while the antenna is within millimetres of sensitive devices or connections.

Some authorities are careful to stipulate the electromagnetic-interference tolerance requirements for equipment, but as a guide the suppliers should be able to guarantee immunity within the usual communications frequency bands (27 MHz, 68–88 MHz, 100–108 MHz, 138–174 MHz and 420–470 MHz). Field strengths of up to 20 V/m may be encountered in these bands, but it is wasteful to overdo the protection: the authority responsible for controlling the overall plant should clearly define the immunity levels that the equipment must possess, and the vendors should be able to demonstrate that their products comply with these requirements.

Apart from radiofrequency interference generated within the plant, consideration should also be given to external sources of interference

EQUIPMENT CONSIDERATIONS 131

(such as nearby broadcast or communications transmitters). If such a situation is encountered, extreme caution is advised: enclosing equipment within a Faraday cage will not guarantee immunity and in fact may make things worse.

6.2.3 Temperature

The heat dissipated within the plant is enough to cause ambient temperatures to rise to high levels, especially when air-conditioning systems break down. At the other extreme, a plant starting up from cold may well be doing so because it has been totally isolated from the power transmission system by, for example, heavy snowstorms—and in this case the equipment ambient temperature may fall to very low levels. The electronic control systems must continue to provide safe and efficient start-up, operation and shut-down of the plant under all such conditions.

Another factor that must be taken into consideration in assessing the temperature environment for the electronic cubicles is the heat dissipation of the equipment itself. Even with careful planning of the thermal dissipation characteristics of the equipment—with printed circuit boards mounted vertically, efficient heatsinks and clear heat-flow paths past critical components—it is possible to encounter temperature rises of up to 12°C within a cubicle. What this means is that an adequate margin (usually 15°C) has to be added to the specified cubicle external temperature and the electronic components rated to work within the resultant limit without any degradation of life or performance.

Various authorities specify different ambient temperature conditions for the electronic equipment, and these are classified according to location—but a common upper limit specification is +40°C for equipment rooms and control rooms, implying that the components must be rated at +55°C. In areas adjacent to the plant, outside ambient temperatures of +55°C are often specified, so cubicle-mounted components in such areas must be rated at +70°C. Adjacent to hot plant, temperatures of up to +85°C may be encountered (though it is unusual to find enclosed cubicles in such areas, and so it is necessary to allow only for the temperature rise within a single assembly).

These values will be exceeded in special cases. Examples are marine installations, where +55°C is sometimes specified for extreme-service ambient temperature conditions—requiring components to be rated at +70°C.

Fortunately, modern technology has yielded semiconductor and linear components that are capable of operating at such temperatures (and, incidentally, generating less heat themselves to exacerbate the situation), but it is still worth checking the temperature rating of systems—especially in refurbishment projects where rigid specifications may not have been generated, as would be the case for new installations.

6.2.4 Vibration

Although the mountings of the turbo-alternator and the coal mills will be designed to isolate them mechanically from the surrounding plant, some vibration will inevitably break through, and plant-mounted equipment may well be subjected to high levels of continuous low-frequency vibration.

Once again, the purchaser should be careful to define the vibration levels that the electronic systems must be capable of surviving, and the vendors should be able to show that their equipment complies with the requirements.

6.2.5 Impact, Shock, Rough Usage and Earthquake

The power-plant environment is by no means a gentle one, especially during the stages of initial construction. Many beautiful gleaming consoles have been damaged by coming into violent contact with passing cranes or heavily booted feet! Also, the control systems must continue to protect the plant during extreme weather conditions including (in some areas) typhoons or earthquakes.

6.2.6 Dust and Dirt

Construction of a power plant is a major civil-engineering operation during which significant levels of atmospheric dust will inevitably be encountered. This may include cement, coal dust and metallic particles, as well as airborne traces of chemicals from various operations carried out at the time. When operational, there is a risk that coal dust may be encountered in the ostensibly air-conditioned equipment rooms or control rooms of even the best-managed coal-fired power stations. (Compare this environment with the almost clinical requirements defined in the operational specifications of items such as disk drives!)

6.2.7 Cabling, Shielding and Grounding

Fault studies carried out by Britain's Central Electricity Generating Board[1] have shown that short-duration voltage transients of up to 1 kV can be encountered *between different parts of a power-station earth mesh*. This is within configurations governed by this authority's extremely stringent rules and codes of practice, and therefore similar or worse conditions must be expected to be encountered elsewhere.

Safety earthing of the equipment is mandatory, but this does not yield satisfactory grounding for electronic systems and equipment. The only really satisfactory system is for each electronic cubicle to be provided with an insulated earth bar—a terminal strip providing a clean earth for all electronic signals in that locality. This *must* be isolated from the surrounding metalwork, which may well experience voltage transients (for the reasons stated above)—and to avoid earth loops[2] it must be linked by a single-point connection to a clean electronic-system earth provided expressly for that purpose. In practice, commercial computer systems are often constructed in such a way that meeting this criterion is very difficult.

Cables must obviously be extremely robust, and most authorities demand the use of steel-wire armouring. Electrical contractors who are responsible for installing such cables are often unaware of the special requirements of electronic-system earthing, and in fact are often trained to ground all armouring wherever possible, in order to protect personnel against electrical shock. It is not always easy to persuade these people to install and make proper use of items such as insulated cable glands. Supervising installers on a huge, complex and busy site—to ensure that correct practices are followed—is also problematical.

Nevertheless, all these precautions must be followed, because it is very difficult indeed to trace a faulty (perhaps duplicated) ground connection after the system has been completed.

6.3 THE NEED FOR COMPREHENSIVE SPECIFICATIONS

The foregoing should have demonstrated the need for a clear definition of the requirements governing the electronic systems installed on a power plant. Such a specification assures not only that the systems will work reliably and safely, but also that competing vendors are assessed on a fair and equable basis. Without such a specification, very divergent bids will be received for tackling a given task, and it will then become difficult to evaluate the alternatives.

It should also be remembered that the task of planning, selecting, purchasing, installing and commissioning a power-plant boiler control installation is a very large-scale undertaking which may occupy a timeframe of several years. (In extreme cases, equipment has actually become obsolete during the progress of the contract, before it could be commissioned!) It is also a process that calls for careful integration with other operations—for example, providing access and space for equipment and people, ensuring that cableways are available, that cut-outs are provided in floors for cables and that environmental requirements are met by the time the equipment is delivered. It is vital, therefore, that the whole process is very carefully planned, operated and monitored—and this again comes back to specifying all the requirements very carefully at the outset.

The overall requirements that need to be covered by the specification are very extensive, ideally covering topics such as the following:

Environmental

- temperature
- dust and dirt
- humidity
- shock
- vibration
- seismic shock
- electrical/electromagnetic compatibility

Administrative

- contract engineering and support
- systems engineering and support
- manufacturing facilities
- training facilities
- progress chasing
- subcontractor assessment
- freight/shipping facilities
- quality assurance
- facilities/qualification
- availability of source code (in case the vendor ceases to trade)
- ability to accept performance bonds

Functional

- system performance
- software structure
- on-line diagnostic facilities
- execution speeds
- failure detection and alarm facilities
- fault tolerance
- effects of failure
- manual back-up facilities
- operator facilities
- communications protocols
- I/O count
- displays (speed/resolution, etc.)
- cabling arrangements (glanding, top/bottom entry, etc.)
- facilities for communication with other systems
- system availability

Quality assurance

- production testing
- factory test (of the system)
- heat-soak testing (of items and systems)
- incoming component test
- availability of remote diagnostics support
- site supervision
- safety procedures
- post-delivery reliability tracking

Service/support

- installation
- commissioning
- training
- documentation (descriptions and diagrams)
- instruction/repair manuals
- spares (location of and duration of guaranteed availability)

At the outset of a project (to build a new system *or* to refurbish an old one) the above lists should be examined to see what is relevant, to delete those points that are not needed and to add any others arising as a result of special circumstances or omissions. This will lead to the production of a specific list of requirements for the project—which at this stage will be outline only and will therefore need to be expanded. For example, if it has been decided that vibration is a critical factor on a given application, a detailed vibration specification will need to be defined.

Once the requirements have been detailed, they should be applied equally to all the bidders for a project, and it will be necessary to ensure that the system that is finally selected does actually comply with the specified requirements and that the vendor really does provide the defined services. This process of assessing the vendor and the system against the requirements should be an on-going operation, from the placement of the initial order, through the various stages of executing the contract, during the installation and commissioning phases, and throughout the subsequent life of the system.

As the last phase of the commissioning operation, the vendor should be required to demonstrate the performance of the systems. This will be related to load changes—a matter that raises a set of additional comments as discussed below.

6.4 PERFORMANCE TESTING

Basically, if a boiler's load is held absolutely constant and if the fuel quality and the air and feed-water temperatures are steady, the entire plant can be put under manual control and all pressures, temperatures and flows will be held constant without the need for any adjustment.

Practically, of course, this situation can never occur—which is, after all, the reason why automatic control systems are needed in the first place. The point is that the effectiveness of the automatic control systems can be measured by the variations they permit in the controlled parameters when changes occur in the boiler load or the quality of the coal, feed water or air.

The performance of a control system is of course governed by the controllability of the plant itself, and one indication of what is limiting the response is whether the controlled element (valve, damper, vanes, etc.) has reached the limits of its travel—on the basis that, for example, 'the temperature is too high but the sprays are flat out, so the sprays are not big enough'. This is a fallacious criterion, however, because a sluggish control system, responding too slowly to a detected change in the plant, may well result in the controlled element reaching its end stop, whereas a good system would have reacted earlier and the element would never have entered that limiting condition.

The performance of the control loops should therefore be evaluated via a series of carefully monitored tests, and these should be meticulously planned. For example, no test should involve changing the number of operating auxiliaries (unless, of course, the system is itself designed to achieve that objective). On a coal-burning plant, therefore, the test of the system's response to a load change should be within $1/N$ times the boiler MCR, where N is the *minimum* number of mills needed to meet 100 per cent MCR. (The reason for specifying the minimum number of mills here is that, if the plant has been conservatively designed, it may be possible to meet MCR with fewer mills than have been provided. If the full complement of mills is used, the turndown assumed for each of these will be less than that which might reasonably be expected.)

The performance that can be expected from a system will need to be agreed between the control system vendor, the boiler maker and the end user, and will clearly vary from installation to installation. However, as a rough guide, the following performance targets have been achieved by 660 MW coal-fired boilers:

Turndown	75 per cent MCR
Maximum rate of change of load	5 per cent MCR/minute

Within these limits, the performance shown in Table 6.1 can be expected (and has been attained by operating plants). The performance does, of course, vary over the load range since the plant response becomes more sluggish at the lower loads where steam, water and gas flow rates are slower. This is reflected in Table 6.1, where the allowable deviations in each parameter are given for loads on either side of an arbitrary breakpoint of 50 per cent MCR.

Table 6.1 Possible expected performance.

Parameter	Allowable deviation <50 per cent MCR	>50 per cent MCR
Drum level (mm)	± 35.0	± 30.0
Furnace pressure (mbar)	± 0.5	± 0.5
Steam pressure (bar)	± 4.0	± 3.0
Steam temperature (°C)	± 20.0	± 8.0

The figures for temperature relate to both superheated and reheated steam.

Obviously, demonstrating that the control systems meet these criteria requires the plant to be made available for testing. This can be a problem, especially when economic considerations demand that every effort must be made to generate as many megawatts as possible for as long as possible.

Performance-related terms may well be included in control-system vendors' contracts, where (typically) 5 per cent of the total contract value is held as a retention until the performance of the system has been demonstrated. It is clearly unfair to apply this penalty when the plant is not made available for testing, but if this fact is recognized by all parties when the initial contract is defined, the appropriate provisions can be made.

Penalties should also be defined for late deliveries and for cases where the system does not meet the requirements. This is an important matter, since lateness or poor performance impose cost-burdens on the plant owners that the original system was designed to reduce or eliminate.

6.5 CONTROL-ROOM LAYOUT

The control room is often the showpiece of the power plant: a place so special that special observation galleries are sometimes provided, allowing visitors to watch the operations being carried out in the 'nerve centre' of the plant.

138 BOILER CONTROL SYSTEMS

It is indeed the nerve centre, since the task of operating a power station requires unique knowledge, confidence and presence of mind. The control systems are there to help the operator as much as possible and to give them information on the plant so that they can take any action that may be necessary to adjust its operation.

The control facilities must be comprehensive, the systems easy to use and reliable, and the environment quiet and conducive to thought. Much thought has been given to the 'human factors' or ergonomic aspects of control-desk design, and that subject is covered very comprehensively in specialist textbooks. Here, we shall concentrate on a few of these aspects as they relate to boiler operations.

The factors to be considered in the location and design of the control room can be grouped into two broad categories:

Safety the control room must provide essential operational facilities, and must do so under all conditions of use and at all times —including, ultimately, during emergencies such as fires or earthquakes.

Accessibility the operators should preferably be able (at least) to see the important parts of the plant and possibly to reach critical functions in the case of emergencies.

In many developed countries these factors are well recognized, and provision is made for regulating the location, layout and design of control rooms and the facilities available within them.[3] In summary, however, the rooms should be located on the least-dangerous side of the plant and should themselves include good provision for escape.

Other aspects that need to be considered are as follows:

Lighting Overhead lighting should be adequate for reading documents but should not cause glare or result in distracting reflections being visible on the screens of monitors. In addition, emergency lighting will be needed. In planning the lighting installation, consideration should be given to access to fittings for maintenance purposes: such operations should not interfere with activities involved with controlling the plant.

Duplication Where two or more units are controlled from one room, the desks of these should be identical—*not mirror images* of each other. (Failure to observe this requirement can lead to confusion when an operator moves from one unit to another, and may cause hazards.)

Communications The unit operator must be provided with telephone communication links to various areas of the plant and to the load dispatcher (or grid control). Public address facilities are also needed, but

EQUIPMENT CONSIDERATIONS 139

are difficult to arrange in the high-noise environment of the plant, and therefore paging systems are usually preferred.

Safety People do strange things in power stations, and as far as possible the control-room design should be such that accidental initiation of a function cannot occur. Locating critical control functions in the form of unshrouded push-buttons near the front edge of a desk section is an invitation for somebody to sit on the button and accidentally initiate the relevant action. Control functions accessed via the screens of monitors (touch-screens or cursor-controlled from trackball, joystick or mouse) should require confirmation from a separate location before they are executed—if only to prevent cleaners or visitors from accidentally taking over the control of the plant. Hard-wired control facilities for emergency use are of little value unless hard-wired indications are also provided to allow the operator safely to operate or shut down the plant when the computer system has failed.

Modifications The control systems and facilities will almost inevitably need to be modified over the lifetime of the plant. 'Soft' controls (that is, those accessed via the displays on monitors) simplify this task, but— bearing in mind the need for boilers to have 'hard' manual control facilities—thought should be applied to the methods of altering the physical arrangement of switches and indicators on the desks. The use of matrix desks (as shown in Fig. 3.32) is of great value in this respect.

The use of multiplexed operator facilities for the control of mills has already been mentioned, but, where these cannot be provided, careful thought should be applied to the methods of providing the operator with the information needed to control the plant, without saturating his or her brain with too much data.

NOTES

[1] Now replaced by two separate authorities: National Power and PowerGen.
[2] The powerful magnetic fields that may always be present in a power station, and the fields arising during generator faults, will induce circulating currents in any continuous conducting loop. These will result in the appearance on earth lines of stray voltages, which are comparable with the low-level signals used by the electronic devices.
[3] Some of these are listed in the publication *Control Rooms for Conventional Power Stations: Planning Instructions*, E65653–P0109–A100–A1–7600, Siemens AG, ZVW85, Fürth-Bislohe, Germany.

APPENDIX
ONE
CONVERSION FACTORS AND OTHER USEFUL INFORMATION

CONVERSION FACTORS

Data are given correct to three significant figures or four decimal places, whichever is appropriate.

Length

	in	mm	cm	m	ft
To convert inches to	× 1	× 25.4	× 2.54	× 0.0254	× 0.0833
To convert millimetres to	× 0.0394	× 1	× 0.1	× 0.0010	× 0.0033
To convert centimetres to	× 0.3937	× 10	× 1	× 0.1	× 0.0328
To convert metres to	× 39.4	× 1000	× 100	× 1	× 3.2808
To convert feet to	× 12	× 305	× 30.5	× 0.3048	× 1

Area

	in^2	mm^2	cm^2	m^2	$feet^2$
To convert square inches to	× 1	× 645	× 6.45	$× 6.45 × 10^{-4}$	× 0.0069
To convert square millimetres to	$× 1.55 × 10^{-3}$	× 1	× 0.01	$× 10^{-6}$	$× 1.08 × 10^{-5}$
To convert square centimetres to	× 0.155	× 100	× 1	$× 10^{-4}$	$× 1.08 × 10^{-3}$
To convert square metres to	× 1550	$× 10^6$	$× 10^4$	× 1	× 10.8
To convert square feet to	× 144	× 92900	× 929	× 0.0929	× 1

1 hectare (ha) = 10^4 m^2 = 2.471 acres
1 acre = 4047 m^2

Volume

	in³	mm³	cm³	m³	ft³
To convert cubic inches to	× 1	× 16 400	× 16.4	× 1.64 × 10⁻⁵	× 5.79 × 10⁻⁴
To convert cubic millimetres to	× 6.10 × 10⁻⁵	× 1	× 0.001	× 10⁻⁹	× 3.53 × 10⁻⁸
To convert cubic centimetres to	× 0.0610	× 1000	× 1	× 10⁻⁶	× 3.53 × 10⁻⁵
To convert cubic metres to	× 6.10 × 10⁴	× 10⁹	× 10⁶	× 1	× 35.3
To convert cubic feet to	× 1728	× 2.83 × 10⁷	× 28 300	× 0.0283	× 1

Force/Weight/Mass

	lb	kg	tonne	N
To convert pounds to	× 1	× 0.454	× 4.54 × 10⁻⁴	× 4.45
To convert kilograms to	× 2.21	× 1	× 10⁻³	× 9.81
To convert tonnes to	× 2205	× 1000	× 1	× 9806
To convert newtons to	× 0.2248	× 0.102	× 10⁻⁴	× 1

Low Pressure

	in wg	in Hg	mm wg	mbar	kPa
To convert inches (water gauge) to	× 1	× 0.0734	× 25.4	× 2.49	× 0.2487
To convert inches of mercury to	× 13.62	× 1	× 346	× 33.9	× 3.39
To convert millimetres (water gauge) to	× 0.0394	× 0.0029	× 1	× 0.0979	× 0.0098
To convert millibars to	× 0.4021	× 0.0295	× 10.21	× 1	× 0.1
To convert kilopascals to	× 4.021	× 0.2953	× 102	× 10	× 1

Pressure

	psi	kg/cm²	bar	kPa	MPa	atm
To convert pounds per square inch to	× 1	× 0.070	× 0.069	× 6.9	× 6.9 × 10⁻³	× 0.068
To convert kilograms per square centimetre to	× 14.2	× 1	× 0.9807	× 98.1	× 0.0981	× 0.968
To convert bars to	× 14.5	× 1.0197	× 1	× 100	× 0.1	× 0.987
To convert kilopascals to	× 0.145	× 0.0102	× 0.01	× 1	× 0.001	× 0.010
To convert megapascals to	× 145	× 10.2	× 10	× 1000	× 1	× 10
To convert atmospheres to	× 14.7	× 1.0332	× 1.0133	× 101	× 0.1013	× 1

Flow

	lb/h	kg/h	kg/s	ton/h	tonne/h
To convert pounds per hour to	× 1	× 0.4536	× 0.00013	× 0.00045	× 0.00045
To convert kilograms per hour to	× 2.21	× 1	× 0.00028	× 0.00098	× 0.001
To convert kilograms per second to	× 7940	× 3600	× 1	× 3.54	× 3.6
To convert tons per hour to	× 2240	× 1016	× 0.2822	× 1	× 1.0159
To convert tonnes per hour to	× 2205	× 1000	× 0.2778	× 0.9844	× 1

Heat/Energy/Power

	therm	J	kW h	kcal	Btu
To convert therms to	× 1	× 1.06 × 10^8	× 29.4	× 2.52 × 10^4	× 10^5
To convert joules to	× 9.5 × 10^{-9}	× 1	× 2.78 × 10^{-7}	× 2.38 × 10^{-4}	× 9.5 × 10^{-4}
To convert kilowatt-hours to	× 3.4 × 10^{-2}	× 3.6 × 10^6	× 1	× 860	× 3.4 × 10^3
To convert kilocalories to	× 4 × 10^{-5}	× 4200	× 1.16 × 10^{-3}	× 1	× 3.97
To convert British thermal units to	× 10^{-5}	× 1055	× 2.94 × 10^{-4}	× 0.252	× 1

1 joule (J) = 1 watt second (W s)
1 horsepower (HP) = 33 000 ft lb/min = 0.746 kW
1 metric horsepower = 32 550 ft lb/min = 0.735 kW

OTHER CONVERSIONS/DATA

1 m^3 of water at 4°C (max. density) = 1000 kg
1 m^3 air at NTP = 1.293 kg
NTP = 1 atm and 0°C
STP = 1 atm and 60°F (15.5°C)
1 Imperial gallon = 4.546 litres
1 US gallon = 3.785 litres

CONVERSION FACTORS 143

THERMOCOUPLE CHARACTERISTICS

The table below summarizes some of the important characteristics of the two types of thermocouple that are most commonly encountered in power-plant boiler applications.

Conductor combination (positive wire named first)	Type	Usual temperature range; continuous operation (°C)	Method of identifying conductors in the field
Iron/constantan (Iron/cupro-nickel)	J	0–800	Positive wire is magnetic, and rusts
T1/T2 (Nickel-chromium/ nickel-aluminium) (Chromel/alumel)	K	0–1100	Negative wire is slightly magnetic

RESISTOR COLOUR CODE*

Colour	First digit	Second digit	Multiplier digit	Tolerance (%)	Temperature Coefficient (ppm/°C)
Black	0	0	× 1		200
Brown	1	1	× 10	1	100
Red	2	2	× 100	2	50
Orange	3	3	× 1000		15
Yellow	4	4	× 10 000		25
Green	5	5	× 100 000	0.5	
Blue	6	6	× 1 000 000	0.25	10
Violet	7	7	× 10 000 000	0.1	5
Grey	8	8			1
White	9	9			
Gold				5	
Silver				10	
None				20	

*For examples see page 144.

Examples

Band 1	Band 2	Band 3	Band 4	Band 5	Value, etc.
Brown	Green	Black	Gold	Brown	15 Ω ± 5%, 100 ppm/°C
Orange	White	Gold	Brown		3.9 Ω ± 1%
Green	Blue	Yellow	Red		560 kΩ ± 2%
Yellow	Violet	Green	Silver		4.7 MΩ ± 10%

Note that, with some resistor types the values are indicated by the first four coloured bands, with fifth and (sometimes) sixth bands adding the tolerance and temperature coefficients. In these cases the first three bands give the numerical value, the fourth the multiplier.

APPENDIX
TWO
SYMBOLS USED ON CONTROL DIAGRAMS

Transmitters

- (FT) Flow
- (LT) Level
- (PT) Pressure

Controllers

- (QT) Misc. variable (identified alongside)
- (TT) Temperature

- (FIC) Flow
- (LIC) Level
- (PIC) Pressure

- (QIC) Misc. variable (identified alongside)
- (TIC) Temperature
- [PID] Three-term controller

Miscellaneous control functions/devices

- (f(t)) Time function
- (f(x)) Function (Output = f(x) X Input)
- (H) Manual signal (on desk)
- (H/A) Hand/auto station (auto/manual station)
- Modulating control valve
- Switch

- (>) Maximum selector
- (<) Minimum selector
- (><) Max./min. limit
- (X) Multiplier
- (+) Addition
- (±) Add/subtract

Plant

- Pump or fan
- Generator

Electrical

- X Circuit-breaker
- Transformer
- Current source

145

APPENDIX
Three
MULTILINGUAL GLOSSARY OF TERMS

English	German	French
Acid cleaning	Säurebeizung	Nettoyage avec acide
Actuator	Betätigungsvorrichtung	Servo-moteur
Air	Luft	Air
Air flow	Luftmenge	Débit d'air
Air heater	Luftvorwärmer	Réchauffeur d'air
Air supply	Frischluftzufuhr	Apport d'air, amenée d'air, arrivée d'air
Alarm annunciation	Gefahrenmeldung	Annonce de danger
Alternating current	Wechselstrom	Courant alternatif
Ambient temperature	Umgebungstemperatur	Température ambiante
Ash	Asche	Cendres
Attemperation control	Einspritzregelung	Désurchauffe par injection
Attemperator	Heißdampfkühler	Désurchauffeur
Audible alarm	Alarmgerät	Alarme acoustique
Automatic boiler control equipment	Kesselregelungsanlage, automatische	(Dispositifs de régulation automatique de chaudière
Automatic control	Regelung, automatische	Réglage automatique
Back-pressure turbine	Gegendruckturbine	Turbine à contre-pression
Balanced draught	Ausgeglichener Zug	Tirage équilibré
Ball mill	Kugel-mühle	Broyeur à boulets
Benson boiler, once-through boiler	Benson-kessel	Chaudière monotubulaire, chaudière Benson
Bled steam feed-water heating	Anzapfdampf-Speisewasservorwärmung	Réchauffage d'eau d'alimentation par vapeur de soutirage
Blow-down	Abschlämmung	Purge de dèconcentration
Boiler control console	Kesselpult	Pupitre de contrôle de la chaudière
Boiler control room	Kesselwarte	Salle de commande principale de la chaudière

MULTILINGUAL GLOSSARY 147

English	German	French
Boiler drum	Kesseltrommel	Réservoir de chaudière ballon de chaudière
Boiler feed pump	Kesselspeisepumpe	Pompe alimentaire de chaudière
Boiler start-up	Kesselinbetriebnahme	Mise en route (ou: service) de la chaudière
Booster fan	Zusatzgebläse	Ventilateur auxiliaire surpresseur
Burner	Brenner	Brûleur
Bypass	Umgehungszug	Passage en by-pass
Bypass valve	Umleitventil	Vanne de by-pass
Cable	Kabel	Câble
Cable conduit	Kabelschutzrohr	Conduit protecteur de câble
Calorific value	Heizwert	Pourvoir calorifique
Carbon dioxide	Kohlendioxid	Gaz carbonique CO_2
Carbon monoxide	Kohlenoxid	Oxyde de carbone
Central control room	Leitstand, Hauptwarte	Salle de commande principale
Centrifugal pump	Kreiselpumpe	Pompe centrifuge
Chain grate	Kettenrost	Grille à barreaux articulés
Check valve	Prüfventil	Vanne de contrôle
Chimney draught	Schornsteinzug	Tirage per cheminée
Chimney emission, chimney discharge	Schornsteinauswurf	Échappement à la cheminée (panache)
Coal	Kohle	Charbon
Coal bunker	Kohlenbunker	Trémie à charbon
Coal feeder	Kohlenaufgabe	Alimentation de charbon
Coal mill	Kohlenmahlanlage	Installation de broyage de charbon
Column of water (in WG)	Wassersäule (mm WS)	Colonne d'eau (mm CE)
Combined firing, multiple-fuel firing	Kombinierte Feuerung	Chauffe par combustibles multiples, chauffe par combustibles mixte
Combustion	Verbrennung	Combustion
Combustion chamber, furnace	Feuerraum	Chambre de combustion
Combustion control	Verbrennungsregelung	Réglage de la combustion
Computer, data processing plant	Datenverarbeitungsanlage	Système de traitement de l'information
Computerized control	Regelung, durch Elektronenrechner	Régulation par calculatrice
Condenser	Verdichter	Compresseur, condenseur
Control	Steuerung	Commande
Control desk	Steuerpult	Pupitre de commande
Control device, regulator	Regelvorrichtung	Dispositif de réglage (ou: de régulation)
Control diagram	Regelschema	Schéma de réglage, schéma de régulation
Control loop	Regelkreis	Circuit de réglage
Control room	Meßerte	Salle de contrôle
Cooling tower	Kühlturm	Tour de refroidissement

English	German	French
Cooling water	Kühlwasser	Eau de refroidissement
CO_2 content	CO_2-Gehalt	Teneur en gaz carbonique
Critical pressure	Druck, kritischer	Pression critique
Current	Strom	Courant (électrique)
Current transformer (meter)	Stromwandler (Messinstrument)	Transformateur de courant (électrique)
Damper for draught regulation	Zugsperre	Registre de tirage
Desulphurization	Entschwefelung	Désulfuration
Desuperheater	Kühler, nachgeschalteter	Désurchauffeur à la sortie du surchauffeur
Deviation in regulation, deviation in control	Regelabweichung	Écart de réglage
Dew point	Taupunkt	Point de rosée
Diaphragm	Membran	Diaphragme, membrane
Diaphragm (type) valve	Membran-ventil	Soupape à diaphragme
Differential draught	Differenz-zug	Tirage différential
Differential pressure regulator	Differenzdruckregler	Régulateur de pression différentielle
District heating plant	Fernheizungsanlage	Installation de chauffage à distance, installation de chauffage urbain
Draught loss	Zugverlust	Perte de charge dans le tirage, perte de tirage
Drum	Trommel	Réservoir, ballon
Earth connection (el.)	Erdschluß (el.)	Prise de terre (él.)
Efficiency	Wirkungsgrad	Rendement
Electric resistance thermometer	Widerstandsthermometer	Thermomètre à résistance
Electronic computer	Elektronische Rechenanlagen	Ordinateurs
Electronic control	Elektronische Regelung	Régulation électronique
Electropneumatic control	Elektropneumatische Regelung	Régulation électropneumatique
Electrostatic precipitator	Elektrofilter	Dépoussiéreur électrostatique, électrofiltre
Emergency stop valve	Notabsperrschieber	Vanne d'arrêt d'urgence
Evaporation	Verdunstung	Évaporation
Excess air	Luftüberschuß	Excès d'air
Fan	Gebläse	Ventilateur
Feedback	Rückkopplung	Contre-coup, choc en retour
Feed valve	Speiseventil	Soupape d'alimentation
Feed water	Speisewasser	Eau d'alimentation
Feed-water heating	Speisewasservorwärmung	Réchauffage d'eau d'alimentation
Feed-water regulator	Speisewasserregler	Régulateur d'alimentation
Feeder	Zuteiler	Alimentateur, distributeur
Final-stage superheater	Endüberhitzer	Surchauffeur de sortie, surchauffer final
First-stage superheater	Überhitzerstufe	Surchauffeur primaire

MULTILINGUAL GLOSSARY 149

English	German	French
Flame	Flamme	Flamme
Flame monitor	Flammenwächter	Contrôleur de flamme (sécurité de défaut de flamme)
Flange	Bördel	Bord tombé, collerette rabattue, collet battu
Flow	Fluß, Strömung	Flux, écoulement
Flow meter	Durchflußmesser	Débitmètre
Flue	Fuchs (Rauchgaskanal)	Carneau
Flue gas	Rauchgas	Fumées, gaz
Flue-gas analyser	Rauchgasprüfgerät	Analyseur de fumées
Fluidized-bed combustion	Wirbelschichtverbrennung	Combustion en lit fluidisé
Forced-draught fan	Unterwindgebläse	Ventilateur de soufflage sous grille
Frequency	Frequenz	Fréquence
Fuel	Brennstoff	Combustible (subst.)
Fuel/air ratio	Brennstoff-Luft-Verhältnis	Rapport air-combustible
Fuel oil	(Brennstoff) Heizöl	Mazout, fuel (-oil)
Fuel-oil heater	Ölvorwärmer, Heizölvorwärmer	Réchauffer de mazout
Furnace control	Feuerungsregler	Régulateur de chauffe
Gas analyser	Gasanalysengerät	Analyseur de gaz
Gas flow	Gasströmung	Flux de gaz
Gas turbine	Gasturbine	Turbine à gaz
Gate valve	Schieber	Vanne, registre, tiroir
Gauge glass, inspection glass	Schauglas	Glace (ou: verre) de niveau, verre de regard
Generator	Generator	Générateur, gazogène
Ground terminal (US), earth terminal (UK)	Erdungsklemme	Borne de mise à la terre
Handwheel	Handrad	Volant
Head loss characteristics (for pumps and fans)	Q-H-Kurve (bei Pumpen und Gebläsen)	Courbe caractéristique débit-pression (pompes et ventilateurs)
Header	Sammler	Collecteur
High–low signal alarm	Alarmsignal HW–NW	Alarme de niveau haut–niveau bas
High	Hoch	Haut
High-pressure preheater	Hochdruckvorwärmer	Réchauffer haute pression, réchauffer HP
High water level	Wasserstand, höchster	Niveau d'eau
Hot well	Warmwasserbehälter	Réservoir à eau chaude
Induced draught	Saugzug	Tirage aspiré
Induced-draught fan (ID fan)	Zugverstärker	Appareil générateur de tirage (ventilateur)
Inlet guide vane (in fans), turning vane	Leitschaufel bei Ventilatoren	Guidage d'entrée (ventilateurs)
Insulation	Isolierung	Isolation

150 BOILER CONTROL SYSTEMS

English	German	French
Interference factor	Störgröße	Grandeur interference perturbatrice
Interlock	Verriegelung	Verrouillage
Interstage attemperator, intermediate attemperator	Kühler, zwischengeschalteter	Désurchauffeur intermédiaire
LP stage of a turbine	Niederdruckteil einer Turbine	Étage basse pression étage BP (d'une turbine)
Lagging	Wärmedämmung	Revêtement calorifuge
Latent heat of vaporization	Latente Verdampfungswärme	Chaleur latente de vaporisation
Level	Niveau	Niveau
Lighting-up burner	Zünd-feuerung	Chauffe d'allumage
Limit switch	Endschalter	(Interreupteur de) fin de course
Live steam, superheated steam	Frischdampf	Vapeur vive
Load	Last	Charge
Load range	Lastbereich	Plage (dans un régime de charge)
Load reduction	Lastabsenkung, Lastabnahme	Réduction de charge
Low water level	Wasserstand, niedrigster	Niveau bas
Low-pressure turbine	Nachschaltturbine	Turbine disposée en série
Lubricant	Schmiermittel	Lubrifiant
Main safety interlock	Hauptsicherungsblockierung	Verrouillage en sécurité principale
Main steam line	Dampfentnahmeleitung	Tuyauterie de départ de vapeur
Manual adjustment, hand adjustment, hand setting	Handeinstellung	Réglage manuel
Maximum continuous rating (MCR)	Dauerleistung, maximal	Marche continue maximale (m.c.max.)
Measurement	Messung	Mesure
Measurement of SO_2	SO_2-Messung	Mesure de SO_2
Measuring and equipment	Meß- und Kontrolleinrichtung	Appareillage de control mesure et de contrôle
Meter	Meßgerät	Appareil de mesure
Metering orifice, disc	Blende (Meßblende)	Diaphragme (de orifice mesure)
Mill, pulverizer	Mühle	Broyeur
Moisture in fuel	Brennstoffeuchtigkeit	Humidité dans le combustible
Multistage	Mehrstufig	A plusieurs étages
Natural draught	Natürlicher Zug	Tirage naturel
Natural gas	Erdgas	Gaz naturel
Nozzle	Mundstück, Düse	Buse, gicleur (Brûleur)
Oil burner	Ölbrenner	Brûler à mazout
On–off control	Ein–Aus-Refelung	Régulation par tout ou rien
Once-through forced-flow boiler, once-through boiler	Zwangdurchlaufkessel	Chaudière à flow boiler, circulation forcée en circuit ouvert

MULTILINGUAL GLOSSARY 151

English	German	French
Part load	Teillast	Charge partielle
Peat	Torf	Tourbe
Pendant superheater	Überhitzer, hängender	Surchauffeur pendentif
pH value	pH-Wert	pH (potentiel d'ions d'hydrogène)
Platen-type	Schottenüberhitzer	Surchauffeur platen
Pollution	Verunreinigung	Pollution, impuretés
Position indicator	Stellungsanzeiger	Indicateur de position
Positioner	Stellungsregler	Régulateur de position
Power factor	Leistungsfaktor	Facteur de puissance
Power station	Kraftwerk	Centrale d'émergie
Precipitator, electrostatic precipitator (for dust separation)	E-Filter (zur Entstaubung)	Dépoussiéreur électrostatique, électrofiltre (dépouissiérage)
Preheater	Vorwärmer	Réchauffeur
Pressure	Druck	Pression
Pressure drop	Druckabfall	Perte de charge
Pressure gauge	Monometer	Manomètre
Pressure-reducing plant	Druckminderanlage	Poste de détente
Pressure vessel	Druckbehälter Druckgefäß	Appareil à pression
Pressurized furnace	Überdruckfeuerung	Foyer sous pression, foyer pressurisé
Primary air	Frischluft	Air primaire
Primary steam temperature	Vorüberhitzung (Dampftemperatur)	Surchauffe primaire
Primary, secondary, tertiary air for furnaces	Erst-, Zweit-, Drittluft für Feuerungen	Air primaire, secondaire, tertiaire pour foyers
Proportional controller	Proportionalregler	Régulateur proportionnel
Pulverized coal	Brennstaub	Charbon pulvérisé
Pump	Pumpe	Pompe
Push-button	Druckknopf	Bouton poussoir
Range of control (or adjustment)	Regelbereich	Plage de réglage
Ratio	Verhältnis	Proportion, relation
Recirculating pump	Umwälzpumpe	Pompe de circulation
Refuse incinerator plant, refuse destructor plant	Abfallverbrennungsanlage	Installation d'incinération des ordures, installation d'incinération des déchets
Regulating valve	Regelventil	Vanne de réglage (ou: de régulation)
Reheat	Zwischenüberhitzung	Resurchauffeur
Reheater	Zwischenüberhitzer	Resurchauffeur
Reset time (controls)	Nachstellzeit (Regler)	Temps de réponse (d'un dispositif de régulation)
Resistance thermometer	Widerstandsthermometer	Thermomètre à résistance
Room temperature	Raumtemperatur	Température ambiante
Rotary air heater, Ljungström air heater	Ljungström-Luvo Luftvorwärmer	Réchauffeur d'air type Ljungström, réchauffeur d'air Ljungström
Safety valve	Sicherheitsventil	Soupape de sûreté

152 BOILER CONTROL SYSTEMS

English	German	French
Saturated steam	Sattdampf	Vapeur saturée
Saturated steam temperature	Sattdampftemperatur	Température de vapeur saturée, température de saturation
Sealing air fan	Sperrluftgebläse	Ventilateur d'air d'éntanchéité
Secondary air	Zweitluft, Sekundärluft	Air secondaire
Secondary superheater	Nachüberhitzer	Surchauffer secondaire
Selector switch, automatic-to-manual selection	Wahlschalter 'Hand–Automatik'	Commutateur 'manuel–automatique'
Shaft horsepower, brake HP	Wellen-PS	Puissance sur l'arbe
Smoke density alarm	Rauchdickealarm	Alarme par opacimmètre
Soot-blower	Rußbläser	Ramoneur, souffleur de suies
Specific heat	Spezifische Wärme	Chaleur spécifique
Spray attemperator	Einspritzküler	Désurchauffeur par injection
Spray valve	Einspritzventil	Vanne d'injection
Spreader stoker	Rost mit Blastisch	Grille avec distributeur pneumatique
Steam	Dampf	Vapeur
Steam atomizer	Dampfdruckzerstäuber, Dampfzerstäuber	Pulvérisation à vapeur, atomiseur à vapeur
Steam demand	Dampfbedarf	Demande de vapeur
Steam flow meter	Dampfmengenmesser Durchflußmesser für Dampf	Débitmètre de vapeur
Steam pressure	Dampfdruck, Dampfspannung	Pression de vapeur, Tension de vapeur
Steam pressure at superheater outlet	Entnahmedruck	Pression (de vapeur) à la sortie du groupe
Steam temperature	Dampftemperatur	Température de vapeur
Steam turbine	Dampfturbine	Turbine à vapeur
Suction	Rückschlagklappe	Clapet de retenue, clapet de non-retour
Suction fan, induced-draught fan (ID fan)	Saugzugventilator, Saugzuggebläse	Ventilateur de tirage aspiré
Sulphur content	Schwefelgehalt	Teneur en soufre
Supercritical pressure	Überkritischer Druck	Pression supercritique
Superheated steam	Heißdampf, Überhitzerdampf	Vapeur surchauffée
Superheated steam temperature	Heißdampftemperatur	Température de surchauffe
Supervisory control equipment, control and monitoring system	Steuer- und Überwachungsausrüstung	Équipement de commande et de contrôle
Switch	Schalter	Interrupteur
Temperature	Temperatur	Température
Thermal stress	Wärmespannung	Contrainte thermique
Thermocouple	Thermoelement	Thermo-couple
Thermocouple cold junction	Kaltlötstelle des Thermometers	Soudure froide (thermo-couple)

MULTILINGUAL GLOSSARY **153**

English	German	French
Thermometer pocket	Thermometertauchhülse	Doigt de gant de thermomètre
Three-element control system	Drei-impulsregelung	Régulation à trois paramètres
Throttle	Drossel	Étranglement, rèduction
Transformer	Transformator	Transformateur
Transmitter	Geber, Übertrager	Transmetteur
Tube mill	Rohrmühle	Broyeur à boulets
Turbine	Turbine	Turbine
Turbine condenser	Turbinenkondensator	Condenseur de turbine
Turn-down ratio	Lastabsenkungsverhältnis	Rapport de rèduction de charge
Unit operation	Blockbetrieb	Fonctionnement par tranches (ou: par unités)
Vacuum	Vakuum	Vide
Valve	Ventil	Soupape, vanne, robinet à soupape
Vane control	Leitschaufelregulierung	Règlage des aubages directrices
Variable-pressure operation	Gleitdruckbetrieb	Marche sous pression variable
Vibration	Schwingung	Vibration
Voltage	(Strom-)spannung	Tension, voltage
Waste heat boiler	Abhitzekessel	Chaudière de récupération
Water level	Wasserstand	Niveau d'eau
Water treatment	Wasseraufbereitung	Préparation de l'eau
Wet steam	Naßdampf	Vapeur humide
Wheel	Rad	Roue

INDEX

Acid rain, 18
Activated Charcoal Adsorption, 28
Adjacent-burner discrimination, in flame monitoring systems, 119
Air ejectors, 14
Air heater, 20–22
Air leakage into combustion chamber, 54, 56
Alternator, 31
Ambient temperature, effects on electronic equipment, 131
Ammonia scrubbing for sulphur removal from flue gases, 28
Anticipatory control, 38
Ash:
   carbon content of, as a combustion indicator, 58
   content, effects of changes, 42
   of coal, 75
Atomic weight, 17
Attemperators:
   control of, in drum boilers, 96–102
   in once-through boilers, 33, 94
   purpose of, reheat, 16
   superheat, 15
   spray-water differential pressure control requirement, 82
   start-up, 100
   (*see also* Desuperheaters)
Auxiliary firing, in combined-cycle plants, 35, 103

Bagasse, 22
Balanced-draught furnaces, 19, 78
Binary systems, 113

Bitumen, 24
Black smoke, 19, 52, 60
Black-out (loss of electric power to control systems), 130
Blowdown, during boiler start-up, 33
Boiler-following, mode of control, 40
Boiler protection logic, 115
Boiler/turbine unit, 6
Booster fans, in FGD plant, 81
Booster pumps, for reheat spray water, 102
Brown-out (reduction of supply voltage to control systems), 130
'Bumpless' transfer, 68
Burner:
   design for NO_x reduction, 18–19, 29
   lightoff—interaction with other sub-systems, 38
   management, 118–122
   typical, 19
Bypass (*see* Turbine bypass)

Cable glands, insulated, 133
Calorific value, effects of changes, 42
Carbon dioxide:
   analysis, as part of carbon-in-ash monitor, 58
   production in combustion process, 16
Carbon-in-ash monitor, 58
Carbon monoxide:
   analysis, 57
   as by-product of combustion, 17, 52
   as control parameter, 56, 57

154

INDEX

Cascade control system, steam temperature, 97
Chain-grates, 24
Characteristics of steam, 7
Characterization, of fuel and air control devices, 59
Charcoal, activated, adsorption, 28
Chemical equations, combustion process, 16–17
Chemical treatment of feed water, 33
Chilling, of superheaters, 8, 100
Classifier:
 characteristics, effect on pulveriser response, 65
 optimization via carbon-in-ash analysis, 58
Clean earth, for electronic systems, 133
CMR (*see* Maximum continuous rating)
Co-generation plants:
 combined cycles in, 37
 control of, 105–107
 principles, 36–37
Combustibles, in fuel, 75
Combustion:
 control, 51–63
 equations, 16–17
 premature, of coal, 65–66
Compensation:
 disturbances, 39
 mills, number in service, 70
 fans, number in service, 74
 feed water, number of auxiliaries in service, 83
Continuous maximum rating (*see* Maximum continuous rating)
Convection, heat transfer to secondary superheater, 14
Co-ordinated unit master, 47–49
CO (*see* Carbon monoxide)
CO_2 (*see* Carbon dioxide)
Coal:
 composition, typical, 17
 firing, 22–24
 fluidized-bed combustion, 24
 mills (pulverizers), control of, 63–67
 weighers, problems with, 75–76
Coal/water mixtures, 77
Cold feed, positive feedback effects in once-through boilers, 90

Combined-cycle plants:
 co-generation, use in, 37
 control, 102–105
 principles, 35–36,
Combustion control:
 cross-limited, 60–62
 parallel, 58–60
Combustor, fluidized-bed boiler, 24
Communications, control room, 138
Condensate:
 return, 13
 system, 31
Condenser, 8, 13, 31
Control room:
 lighting, 138
 location, 139
Controller:
 air flow, 73–77
 combustion, 58–63
 feed-water, 81–96
 furnace pressure, 78–81
 manual control facility, 124–127
 master pressure, 39
 steam temperature, 96–102
Cooling towers, 36
Corrosion, of furnace tube walls, 19
Critical pressure, 33
Cross-limited combustion control:
 in fluidized-bed boilers, 63, 77
 in gas or oil-fired boilers, 60–62
Cycle efficiency (*see* Efficiency)
Cyclic life-expenditure curve of steam turbines, 34–35

DAC, 124, 126–127
Dampers:
 characterization, 59
 control, 20, 113, 126
 fan discharge, 117
 leakage, 56
 mill (pulverizer),
  primary air control, 72
  suction control, 66
  temperature control, 65–66
 reheat temperature control, 102
d.c. blocking, 79
De-aerators, 31
Demineralized water, use in once-through boiler pre-cleaning, 33
Desulphurization of flue-gases:
 furnace-pressure interaction, 81
 processes, 27–28

Desuperheater, 96
  (*see also* Attemperators)
Dew point of flue gases, 22
Digital control systems, 5, 113–114,
  124–127, 132
Digital-to-analogue converter, 124,
  126–127
Disk-drives, environmental
  considerations, 132
Dissolved oxygen in feed water,
  effects of, 31
Distribution of air to burners, 76
Drum:
  boiler,
    feed-water control, 81–87
    principles, 10
  construction, 32
  level:
    controller, 83
    measurement, duplication of, 85
    Hydrastep, 85
    interaction with steam
      temperature, 38
    transmitter, duplication, 85
  pressure and temperature
    compensation, 85
Dry saturated steam, 8
Dust and dirt, effects on electronic
  equipment, 132
Dust emission, due to poor location
  of furnace draught taps, 79
Dust filters, flue-gas, 27

Earth bar, insulated, 133
Earth connection, for electronic
  systems, 132–133
Earth loops, 133, 139*n*
Economizer, 11–13
Efficiency:
  adverse effect of reheat sprays on,
    16
  of boiler, influence of economizer,
    13
  of Rankine cycle, 8
Electrical power, generated, 9
Electrical supplies, 128–130
Electromagnetic compatibility, 130–
  131
Electronic components, temperature
  ratings of, 131
Electrostatic precipitators, 27

Emergency:
  operation of plant in, 126, 139
  trips, 121
Energy release in furnace, 51–52
Entropy, in Rankine cycle, 8
Environment:
  effects of power-plant operations
    on, 17–18
  of electronic equipment, 128–133
Ergonomics:
  of control room, 138
  national preferences in control-desk
    design, 4
Erosion:
  in fluidized-bed boilers, 24
  of thermocouples, in mill
    (pulverizer) applications, 66
Excess air, 52, 77
Excess firing:
  logic, 115
  multiple-fuel considerations, 68, 72
Excitation, alternator, 31
Exhauster, 66–67
External heat exchanger, in MSFB
  boilers, 24–25
Extraction pumps, 31

Failure-response, of cross-limited
  combustion-control system,
  62
Fans:
  forced draught, 19
  gas re-cycling, 16, 102
  induced draught, 19, 21, 78
  primary air, 23–24
Faraday cage, 131
FD fans (*see* Forced-draught fans)
Feed-forward:
  in boiler control generally, 38
  in furnace-pressure loop, 79
Feed heaters, 14, 31
Feed pumps:
  control, 82
  purpose, 13
Feed valves:
  drum boiler, 82–84
  once-through boiler, 88–96
  start-up, 86
    importance of isolation, 86
Feed water:
  control, 82–96
  effects of temperature changes, 42

INDEX

FGD (see Flue-gas, desulphurization)
FGT (see Flue-gas, treatment)
Filters, dust, 27
Fire-tube boilers, 9
First-to-auto transfer in mill control loops, 68–70
Flame scanners, 119
Flash tank, 33, 49, 90, 94
Flow nozzle, for steam-flow measurement, 87
Flue gas:
　desulphurization, 27
　effects on furnace-pressure control, 81
　emissions, with high-sulphur fuels, 77
　treatment (NO_x-reduction process) 28
Fluidized-bed boiler:
　air-flow control, 77–78
　combustion control, 63
　contribution to pollution reduction, 18
　design, 25
　principles, 24
　steam-temperature control in, 101
Forced-draught fans:
　control, 73–77
　purpose of, 19
Fuel/air ratio, 56
Fuel and air mixing, imperfections in, 52
Fuel systems, 22
Furnace:
　influence of FGD system, 81
　pressure control, 79–81
　pressure tappings, 78–79
　protection, 79–80, 115
　purge, 119, 121–122
　suction, 19, 78
　(see also Induced-draught fans)

Gain compensation, for number of mills in service, 70
Gas firing, 22
Gas recycling:
　control dampers, 102
　fans, 11, 16
Gas turbine:
　use in combined-cycle plants, 35
　control of, 102–103
Generator, electrical, 6

Generator faults, 129
Global climatic effects, 18
Governor, turbine, 16, 30, 40–43, 46, 48, 50
Grid:
　control, 138
　frequency disturbance contributed to by turbine-following control, 42
　power connection of generators, 6
Grounding, electronic, 132–133

Hand/auto station:
　manual control considerations, 124–127
　mill master 70–71
Hanging, of sequences, 116
Hard-wired control and indications, 139
Heat dissipation of electronic equipment, 131
Heat losses, in boiler, 52–53
High-pressure (HP) bypass (see Turbine bypass)
High-pressure (HP) stage of turbine, 10
Hoppers, dust, 27
Hot re-start, using turbine bypass valves, 34
HP bypass (see Turbine bypass)
HP stage of turbine, 10
Human factors (see Ergonomics)
'Hunting' of FD fans, 75
Hydrastep, drum-level measurement, 85
Hydrocarbon fuels, 16–17

ID fans (see Induced-draught fans)
Idle burner cooling, 77
Igniters, 119
Impulse, principle in steam turbine, 29
Indicators, hard-wired, 139
Induced-draught fans:
　control of, 78–81
　purpose, 19
Infra-red:
　flame scanners, 119
　pyrometer, use in fluidized-bed boiler control, 78
Interaction of boiler sub-systems, 38

Interlocks:
  bypass valves, 110
  design of, 114–115
Intermediate-pressure stage of
    turbine, 13
IP stage of turbine, 13
IR flame scanners, 119
'Island of automation', concept in
    combined-cycle control, 104
Isolating valves:
  emergency operation, in event of
    control failure, 126
  importance of, for protecting start-
    up feed regulator, 86

Jacking oil, 118
Joystick, computer, 139

Latent heat, 7
Leakage of air into furnace, 56
Lighting of control rooms, 138
Limestone:
  flue-gas desulphurization use, 27
  fluidized-bed boilers,
    control of, 78
    use of, 24
  scrubbing, 27–28
Limit switches, failure of, 116
Ljungström air heater, 21
Load change, effect on drum level, 83
Load dispatcher, 138
Load line, of mills, 64–65
Loss of feed, margin in drum boilers,
    85
Loss of fuel logic, 115
Losses, heat, in boilers, 52–53
Low-NO_x burner design, 19, 29
Low-pressure bypass (*see* Turbine
    bypass)
Low-pressure stage, of turbine, 13
LP bypass (*see* Turbine bypass)

Make-up water, 13, 29
Manual control:
  methods, 124–128
  using stepper-motor drives, 127
Master:
  control, 39–51
  fuel trip, 115, 121
  hand/auto station, for mills, 70
  pressure controller, 41

Matrix desks (*see* Mosaic desks)
Maximum continuous rating (MCR),
    14
Measurement:
  coal flow, 75
  steam flow, 87
Megass, 22
Mill (pulverizer):
  coal, 22
  control, 63–73
  first-to-auto transfer, 68–70
  'load line', 64–65
  logic, 73
  pressurized, control of, 64–66
Mixed-fuel firing, 67–68, 70–72
Moisture:
  content of fuel, effects of changes
    in, 42
  in weight measurement, 75
Monitors, control-room, lighting
    considerations, 138
Mosaic desks, 4–5, 105, 139
Mouse, computer, 139
MSFB (*see* Multi-solids fluidized-bed
    boiler)
Multiple burners, control
    considerations, 62
Multiplex operation of mills, 70
Multi-solids fluidized-bed boiler, 25,
    78, 101

National Fire Prevention Association
    (NFPA), 122
Natural characteristic of boiler,
    temperature, 15
Nitrogen:
  in air, 17
  oxides, 17–18, 28
Noise, electronic, problems when
    using derivative action,
    98
Non-contact attemperators, 15, 96
NO_x (nitrogen oxides), 18, 28
NO_x:
  control, in multi-solids fluidized-
    bed boiler, 78
  production in fluidized-bed boilers,
    24, 29
Nozzle:
  steam-flow measuring, 87
  turbine, 29

INDEX 159

Oil firing
   combined with coal firing, 71–72
   simple, 22
Once-through boiler:
   feed-water control, 87–96
   operation, 9–10, 32
   role of bypass valves in, 34
   start-up, 87–89
   supercritical, 33, 87, 90–96
Opacity of flue-gases, use in
      combustion control, 56–57
Orifice plate, for steam-flow
      measurement, 87
Orimulsion, 24, 77
Over-firing, protection against, 68,
      72, 115
Override:
   manual, 124, 128
   sequence system, 116
Oxygen:
   in combustion air supply, 16, 19,
      56, 58
   content, of flue gases, 22, 57, 77
   dissolved, 31
   flue-gas content, relationship to
      load, 57
   trim, 57

Parallel combustion-control, 58–60
Part-load operation of mills, 54, 56
Penalty clause, in contracts, 137
Pendant superheater, 14
Performance testing, of control
      systems, 136
Piezo-electric igniters, 119
Pilot burners, 119
PLC interlock systems, 113, 115
Pollution:
   atmospheric, 17–19, 28–29, 77
   combined-cycle plant, 36
   dust, 27
   water, 36–37
Power demand:
   master, 39–51
   minimum acceptable step, 39
Power supplies, 128–130
Power supply, uninterruptable, 128
Precipitators, 27
Pre-ignition of coal, 65–66
Pressure-control dampers, 102

Pressure-reduction:
   valve, feeding supercritical boiler
      flash tank, 90–96
   in turbine bypass systems, 34
Pressure-type horizontal tube mill
      (pulverizer), control, 66–67
Pressurized vertical-spindle mills
      (pulverizer), control, 64–66,
Primary-air fans, 23–24
Procedureless bumpless transfer, 64–
      66
Products of combustion, 17–19
Programmable-logic control,
      interlock systems, 113, 115
Proportional band, 98
Protection logic, furnace pressure,
      115
Pulverizer:
   coal, 22
   control of, 63–73
   load-line, 64
   pressurized, control of, 64–66
Pumping and heating, fuel-oil, 22
Pumps:
   extraction, 31
   feed-water, 13–14, 82, 88
   fuel, 22

Q (unit of energy), 18

Radiant superheater, 14
Range connection of boilers and
      turbines, 7
Rankine cycle:
   in combined-cycle plants, 103
   temperature/entropy diagram, 8
Rappers, electrostatic precipitators,
      27
Reaction principle, in steam turbine,
      29
Recuperative air heater, 21
Re-cycle, gas, 11, 16, 102
Redundancy:
   in feed system, 83
   of manual control units, 125, 127
Regenerative air heater, 21
Regulator (*see* Valve)
Reheat steam temperature control,
      101–102
Reheater 16

Reliability:
  of control equipment, 124–133
  of interlocks, 113
Renewable energy plants, 39
Reset wind-up, 98–99
Retention money, in contracts, 137
RFI protection, 130

Sabotage, 115
Safety, 113, 124, 135
Safety earthing, 132–133
Safety valves:
  HP bypass valves used as, 110
  in multiple-fuel plants, 68
Saturation of controller in cascade loops, 98–99
Saturation temperature:
  definition, 7
  steam/water density near critical pressure, 33
  superheat temperature control, operation near, 99
SCR (selective catalytic reduction), 28
Scrubbers, ammonia, 28
Seal-air, of pressurized tube mills, control of, 67
Secondary air, 11, 24
Secondary superheater, 14, 97, 100
Selective catalytic reduction, 28
Selective master, 47
Selective non-catalytic reduction, 28
Sensible heat, 7
Separator, in once-through boilers, 33
Sequence systems:
  design of, 116–118
  in feed-water control loop, 86
  mill (pulverizer), start-up and shut-down, 73
  purpose, 113
Setting leakage, 54, 56
Shock, thermal, 8, 100
Shrinkage, drum water-level effect, 83
Shut-down of mills, 73
Single-element feed-water control, 83
Sliding pressure control, 43–46
Smoke, 52–53, 60
SNR (selective non-catalytic reduction), 28
'Soft' controls, 139
Software, control, 5, 107n

Solids:
  in feed water, 33
  in fuel, 18
Soot-blowing, effect on steam temperature, 97
Spikes, voltage, 129
Spray drying, for sulphur removal from flue gases, 28
Sprays (see Attemperators)
Start-up:
  mills, 73
  once-through boilers, level control, 88–89
  recirculation flow, 89
  vessel, once-through boilers, 10
Station batteries, unsuitability for use with control systems, 130
Steam:
  characteristics of, 7
  flow,
    adjustment of oxygen control set value, 57
    measurement, as indication of fuel flow, 76
    measurement methods, 87
  tables, 7
  temperature control,
    in conventional boilers, 96–101
    in fluidized-bed boilers, 96, 101
    in once-through boilers, 33, 94
    time-constants and transfer functions, 96, 101
    in turbine bypass systems, 34
  turbine (see Turbine, steam)
Steam-flow/air-flow ratio, control of FD fans, 76
Steel-wire armouring, cables, 133
Stepper-motor drives, 4, 127–128
Stoichiometric fuel/air ratio:
  combustion control in conventional boilers, 52
  definition, 9
  fluidized-bed boiler, 78
Stoker, chain-grate, 24
Storage capacity of boiler drums, 85
Stress, in turbines and boilers, 34
Stripping, thermal, 28
Suction-type mills (pulverizers):
  horizontal tube, control of, 66
  vertical-spindle, control of, 66
Sulphur, in fuel, 27
Sulphur dioxide, 17
Sulphuric-acid formation, 22

INDEX

Supercritical:
  boilers,
    feed water control, 90–95
    principles, 33
    pressure, 33
Superheated steam:
  nature of, 8, 10
  temperature control, 96–101
Superheaters, 10, 14
SWA (steel-wire armouring), 133
'Swell', drum water-level effect, 83
Synchronous speed, 16, 31

Tapping point, furnace-pressure
    measurement, 79
Temperature, ambient, of equipment, 131
Tempering air, 65–66
Testing, acceptance, 136–137
Thermal efficiency:
  of co-generation process, 37
  of combined-cycle plants, 36–37
  of Rankine cycle, 8
Thermal shock, 8, 100
Three-element feed-water control, 84
Three-term controller, use in steam-
    temperature control, 98
Time constant:
  reheater, 500 MW unit, 101
  superheater, 500 MW unit, 96
Time-delay, non-linear, control
    element in combustion
    control, 60
Touch-screens, 139
Trackball, 139
Tramp air, 56
Transfer function:
  reheated-steam temperature system
    of 500 MW unit, 101
  superheated-steam temperature
    system of 500 MW unit, 96
Transport time, in coal systems, 75
Turbine, gas (*see* Gas turbine)
Turbine, steam:
  blades, 29–30
  bypass,
    consideration of, in steam flow
      measurement, 87
    control, 109–112
    purpose of, 33–34
    use as safety valves, 110
  governor, 16, 30, 40
  high-pressure (HP) stage, 10

intermediate-pressure (IP) stage, 12
low-pressure (LP) stage, 12
principles, 29–31
rotor, 30
stresses imposed by temperature
    changes, 35
trip, 115
Turbine-following mode of control, 41–43
Turbo-alternator, 29–31
Turndown:
  boiler, 39–40, 136
  mills, 39
Two-element feed-water control, 83
Two-out-of-three logic, in protection
    systems, 115

Ultra-violet flame scanners, 119
Uninterruptable power supply, 128
Unit, boiler/turbine, 6–7
UV flame scanners, 119

Vacuum, condenser, 31
Valve:
  feed-water regulating, 119, 123, 126, 131
  governor, 30
  isolating, 86
  turbine bypass, 33–34, 87, 109–110
  turbine governor, 30
Vanadium, in fuel, effects of, 77
Vanes:
  forced-draught fan, 76
  induced-draught fan, 80
Variable-pressure operation, 43–47
VDUs, lighting considerations, 138
Vibration, 132, 135
Voltage transients, 129, 132

'Walkie-talkies', 130
Waste heat:
  boiler, 35
  dumped to environment, 36–37
Waste products of combustion, 17–18
Waste water, from flue-gas de-
    sulphurization processes, 28
Water, make-up, 13, 29
Water-tube boilers, 9
Weighers, coal, 75
Wet saturated steam, 8
Windbox:
  FD fans feeding, 73
  pressure control, 76–77